Independent Study IS-7 September, 2003

EMERGENCY MANAGEMENT INSTITUTE

A CITIZEN'S GUIDE TO DISASTER ASSISTANCE

FEMA

CONTENTS

FOREWORD

The Federal Emergency Management Agency (FEMA) is the central point of contact within the Federal government for a wide range of emergency management activities, both in peacetime and in time of war. The agency has numerous roles, including coordinating government activities, providing planning assistance, guiding and advising various agencies, and delivering training.

FEMA's training program is delivered through the Emergency Management Institute (EMI) and the National Fire Academy (NFA). These schools are collocated on the National Emergency Training Center (NETC) campus at Emmitsburg, Maryland. NFA is the national focal point for Federal efforts to advance the professional development of fire service personnel engaged in fire prevention and control activities. EMI provides emergency management training to enhance emergency management practices throughout the United States for the full range of potential emergencies.

Both NFA and EMI offer courses, workshops, and seminars on the Emmitsburg campus as well as nationwide through the emergency management training program and State fire service training programs. Although most training activities are directed at State and local government officials with emergency management or fire protection responsibilities, some are provided to private sector and volunteer agency audiences, as well as to the general public. A complete listing of EMI and NFA courses is also available on FEMA's World Wide Web site. The address is http://www.fema.gov.

Independent Study Courses

FEMA's independent study program is one of the delivery channels EMI uses to deploy emergency management training to the general public and to emergency management audiences. The independent study program includes courses in radiological emergency management, the role of the emergency manager, and hazardous materials. Many of these independent study courses are available at FEMA's World Wide Web site.

These independent study courses are geared toward both the general public and persons who have responsibilities for emergency management. All courses are suitable for either individual or group enrollment and are available at no charge. Courses include a final examination. Persons who score 75 percent or better on the final examination are issued a certificate of completion by EMI.

If you desire additional information about these courses, contact your local or State Office of Emergency Management or write to:

FEMA Independent Study Program
Emergency Management Institute
16825 South Seton Avenue
Emmitsburg, MD 21727

INTRODUCTION

A disaster can disrupt our lives with little or no notice, and as the events of September 11, 2001, demonstrated, citizens need to prepare themselves for all types of hazardous events. When disaster does strike, citizens may be caught unprepared for the variety of questions that arise about how to protect themselves, their families, and their property. After the emergency is over, they confront additional crucial needs for information. Where can they get help? What assistance is available?

This independent study course will address these issues and explain the importance of pre-disaster preparedness. It is intended for the general public as well as those in the emergency management community who want to obtain a basic understanding of disaster assistance. It is not the purpose of the course to be an operations manual for disaster assistance programs.

No prior knowledge of the subject is assumed. This course will provide each reader with a foundation of knowledge that will enable him or her to:

◆ Understand what disaster assistance is and what it is not.

◆ Describe how community, State, and Federal governments, as well as voluntary agencies, respond to disasters and how they assist citizens during the recovery phase.

◆ Explain how people obtain assistance in the event of a disaster.

◆ Conduct preparedness activities that help individuals and families survive and recover from a disaster.

Course Overview

This course has five units.

Unit One, Introduction to Disaster Assistance, will provide an overview of disaster assistance and explain the responsibilities of communities, voluntary agencies, State government, and the Federal government in providing relief services. It will also deal with common misconceptions about disaster assistance.

Unit Two, How Communities and States Deal With Disasters, will describe the response and recovery activities that occur at both the local and the State levels in the event of a disaster.

Unit Three, Overview of Federal Assistance, will describe the role of the Federal government in disaster assistance and provide a brief history of Federal disaster relief. The process by which a Presidential disaster declaration is made will be explained, and Federal assistance programs will be discussed.

Unit Four, Federal Disaster Assistance in Action, will describe common sources of information about Federal major disaster assistance and explain how individuals and communities can obtain this assistance.

Unit Five, The Citizen's Role in Disaster Preparedness, will present preparedness activities that help individuals and families prepare to survive a disaster.

How to Complete the Course

This course is designed to be completed at your own pace. You will remember the material best if you do not rush through it. Take the time to study the material and jot down comments: the blank space next to the text is a good place to take notes. Take a break after each unit and give yourself time to think about what you have just read. Then take the short quiz at the end of the unit. Check your answers. If you have missed any questions, go back and review the material again.

The course contains a pretest, five units, two appendices, a glossary, a resource section, and a final examination. You can score the pretest yourself using the pretest answer key on page K-1 and determine how much you know about disaster assistance before you begin studying the course.

The glossary contains definitions of important terms used throughout the course. Consult it frequently to refresh your understanding of how key words are used in the text.

A resource section is included to help you continue learning after completing the course. This section features listings of recommended readings to provide additional information relevant to the course.

The final examination, located at the end of the course booklet, will test the knowledge you have gained from the course. An answer sheet is supplied with the course materials. Mail the completed answer sheet to the address on the form. Your test will be evaluated and results will be mailed to you within a few weeks. If your score is 75 percent or above, a certificate of completion will be mailed to you.

How to Take the Pretest

The following pretest is designed to evaluate your current knowledge of disaster assistance. Read each question and all the possible answers carefully before you mark your answer. There is only one correct answer for each test item. Mark the answer by circling the correct response.

There are 25 questions on the pretest. The test should take you approximately 15 minutes. Find a quiet spot where you will not be interrupted during this time.

After you have checked all your answers using the answer key on page K-1, begin reading Unit One.

PRETEST

The pretest is designed to evaluate your knowledge of disaster assistance. Read each question carefully and select the one answer that best answers the question. Circle the letter corresponding to the answer you have chosen. Complete all the questions without looking at the course materials.

When you have answered all the questions in the pretest, check your answers against the answer key that is provided on page K-1 at the back of the course materials. Your score will be meaningful only if you have answered all the questions before you begin the course.

The pretest should take you approximately 15 minutes to complete. When you have finished, and all of your answers have been checked, begin taking the course.

1. The natural disasters that most frequently result in the loss of lives and property damage are _____.

    a. Earthquakes.
    b. Volcanoes.
    c. Floods.
    d. Landslides.

2. The majority of emergencies are handled by _____.

   a. The local government.
   b. The local government with State assistance.
   c. The local government with State and Federal assistance.
   d. The Federal government.

3. Federal agencies can offer no assistance in a disaster unless there is a Federally declared disaster.

   a. True
   b. False

4. Everyone in a Presidentially declared disaster area is eligible to receive disaster assistance.

   a. True
   b. False

5. Financial aid received from the Federal government following a disaster is intended to address whatever needs the recipient considers most urgent.

   a. True
   b. False

6. Federal disaster assistance can be used to replace any item damaged in a disaster.

   a. True
   b. False

7. Repairing damages caused during an emergency and returning to normal life is called _____.

   a. Planning.
   b. Preparedness.
   c. Response.
   d. Recovery.

8. Local government responsibilities include all of the following: publicizing available assistance, providing situation reports to the State in a serious emergency, and coordinating with emergency management officials from neighboring jurisdictions and the State to supplement disaster response capabilities.

   a. True
   b. False

9. The primary responsibility for helping residents recover from emergencies rests with which level of government?

   a. Local
   b. State
   c. Federal
   d. None

10. An example of an activity needed for short-term recovery would be

    _____.

    a. Making houses habitable so people can return to them.
    b. Considering strategies that would lessen the effects of a similar disaster in the future.
    c. Strengthening building codes.
    d. Relocating damaged structures.

11. The Federal agency responsible for advising the President on whether to declare a major disaster is the

    a. Occupational Safety and Health Administration (OSHA).
    b. Federal Emergency Management Agency (FEMA).
    c. Environmental Protection Agency (EPA).
    d. Department of Transportation (DOT).

12. Which of the following is an example of hazard mitigation?

   a. Elevation of flood-prone structures.
   b. Strengthening of an existing structure to withstand high winds.
   c. Initial implementation of vegetation management for wildfires.
   d. All of the above.

13. A police or fire station damaged in a disaster might be eligible for what form of Federal assistance?

   a. Small business loans.
   b. Individual and Family Grant Program.
   c. Public Assistance.
   d. Hazard Mitigation Grants.

14. Officially, the Federal disaster declaration process begins when the _____ makes a request for a Presidential disaster declaration.

   a. Local Emergency Manager.
   b. FEMA Regional Director.
   c. President.
   d. Governor.

15. The _____ evaluates the Governor's request for a Presidentially declared disaster, then makes a recommendation to the Director of FEMA.

   a. FEMA Regional Director.
   b. President.
   c. Governor.
   d. Local emergency manager.

16. A Preliminary Damage Assessment estimates the extent of the damage, documents it, and helps establish the need for Federal help.

   a. True
   b. False

17. The person responsible for coordinating the overall Federal disaster recovery effort in a Presidentially declared disaster is the

    _____.

    a. Local elected official.
    b. Governor.
    c. Federal Coordinating Officer.
    d. State Coordinating Officer.

18. Disaster victims seeking to apply for Federal disaster assistance following a Presidentially declared disaster could _____.

    a. Visit their local emergency management office.
    b. Visit any Red Cross shelter.
    c. Call FEMA's National Teleregistration Center.
    d. Visit a Disaster Field Office.

19. The unit that coordinates Federal disaster relief and recovery efforts in a Presidentially declared disaster is _____.

    a. The Disaster Recovery Center staff.
    b. The Emergency Response Team.
    c. The emergency management office staff.
    d. The Congressional office staff.

20. In a Presidentially declared disaster, victims can apply for assistance by telephone.

    a. True
    b. False

21. The purpose of hazard mitigation is to _____.

    a. Dispense aid to families.
    b. Replace televisions, refrigerators, and other personal property.
    c. Reduce potential damages from future disasters.
    d. Pay for crop damage.

22. Evacuations occur hundreds of times each year throughout the United States.

    a. True
    b. False

23. If it were necessary to evacuate because of a disaster, 24 hours notice would always be provided to enable evacuees to prepare.

    a. True
    b. False

24. You can reduce the chance of serious loss in your home by
    _____.

    a. Installing a smoke detector on each floor.
    b. Planning alternate escape route.
    c. Purchasing and learning to use an ABC-type fire extinguisher.
    d. All of the above.

25. A family preparedness plan should include a meeting place in the neighborhood where they live for what purpose?

    a. To gather family members after a disaster and verify their safety.
    b. To serve as a site for disaster-related practice sessions.
    c. To provide a location for warning equipment.
    d. To keep extra keys in case they are needed.

Unit One

Introduction To

Disaster Assistance

Those who have never experienced a disaster may find it hard to anticipate all the ways a disaster could affect them. This unit begins with a scenario that dramatizes a major flood; it highlights many of the needs that result from disasters and the actions typically taken to meet them. This scenario describes the disaster's impact on the community, showing actions governments and other agencies could take to respond to such a disaster and to help the community recover from its damage. It also introduces actions that communities and individuals can take to lessen the impact of disasters on homes and businesses.

In this unit, you will learn about:

◆ The purpose of disaster assistance.

◆ Overall responsibilities of local communities, voluntary agencies, Federal, State, and tribal governments.

◆ Common misconceptions about Federal disaster assistance.

Following the scenario is an overview of disaster assistance that provides information on what help might be available at the local, State, Federal, and tribal levels. Finally, this overview discusses the false beliefs some people hold about the role of the Federal government in disasters.

A DISASTER STRIKES

The disaster story you are about to read describes an event that is common in many parts of the United States—a flood. Floods are the natural events that most frequently result in the loss of lives and property damage, claiming an average of 263 lives every year. Although this example deals with the flooding of a river, many of the consequences described could also result from a hurricane, earthquake, or tornado.

The Centerville Flood

Before the flood, rain fell steadily for several days. An unusually wet season had already left the ground saturated and unable to absorb much more rainwater.

Many of Centerville's streets looked more like rivers.

Because of the geographic characteristics of the region and the path of the nearby river, minor flooding has always been common in the area. However, the National Weather Service had provided information that alerted local officials to a more serious possibility—the occurrence of a major flood that had the potential to threaten lives and property and would even necessitate a major evacuation.

Local officials carefully monitored water levels and continued to coordinate frequently with the National Weather Service. As it became apparent that the rising water was indeed creating flood conditions much more threatening than the minor floods of the past, local officials alerted the State emergency management agency. Warnings were issued in the local media advising residents that serious flooding was possible and that they should be prepared to evacuate. When flood waters finally overflowed the river banks, many telephone and electricity lines came down in affected

Outlying communities also experienced flooding.

residential areas. A number of residential streets were so severely flooded that they looked like rivers. Partially submerged vehicles—abandoned by owners seeking safety and higher ground—littered the streets.

Local television and radio stations announced that an emergency was being declared by the local government and that the town's emergency response plan was being activated. Residents of low-lying areas were advised to evacuate and directed to use particular routes to reach designated shelters (operated by the American Red Cross and other voluntary agencies) located a safe distance from the flooded area. Waters were rising so rapidly in the heavy rain that people had to be evacuated from some neighborhoods by boat.

In areas where power lines and phone lines were down, only people with battery-operated televisions or radios could receive the call to evacuate. Many residents could not make phone calls to obtain information; those who could, frequently received busy signals. Some who tried to drive to safety were unfamiliar with which routes could still be traveled and were injured on flooding highways.

Others underestimated the floodwater's power and tried to walk through flooding streets. Because it is possible to be swept away by only a foot of water, additional injuries occurred as a result of this error in judgment.

As waters continued to rise, exposed power lines, structural damage, and floating

debris posed safety hazards. Major transportation routes were at risk. Pieces of damaged buildings and other debris created a dam, causing the flood-waters to spread out yet further into the community. The earthen flood-protection works were weakening.

Many homes, businesses, and important structures such as hospitals and nursing homes were damaged. Medical facilities that were still intact were overwhelmed with people seeking help. Every community in the area experienced some devastating damage from the flood.

Outlying semi-rural areas were damaged as well as Centerville. In these areas, people had a harder time reaching medical assistance because the distance to be traveled through hazardous conditions was much greater. Local government officials again contacted the State emergency management agency. They informed the State that the situation had surpassed the local capability to respond and requested assistance.

In the wake of the flood, many homes were uninhabitable and many businesses required major repairs before they could reopen.

Acting on information provided by local officials, the State declared Centerville a disaster area. As the situation continued to worsen, the State asked regional officials representing the Federal Emergency Management Agency (FEMA) to conduct a preliminary damage assessment. Damage assessment teams (which included representatives of local, State, and Federal agencies) immediately recognized the severity and magnitude of the damage. The flood-protection works had failed. Devastation had by then spread over a number of jurisdictions in an extended geographic area, and many buildings, facilities, and institutions were destroyed or inoperable. The flood had done major damage to businesses that provided essential goods and services such as medicine, food, hardware, and clothing. One school was severely damaged. Hundreds of homes were washed away, destroyed, or rendered uninhabitable by the floodwaters. Many people had lost their property or suffered injuries.

The damage was so severe that it seemed unlikely that local and State governments could meet the disaster-related needs of the community. The Governor consulted with local and State emergency management/ officials, as well as FEMA regional staff. They reviewed damage assessment reports, identified needs, and considered the capacity of available resources. On the basis of this information, the Governor decided to submit a request through FEMA that the President declare the area a major disaster. The President issued the declaration to provide the assistance requested by the Governor.

*Every community is vulnerable to some type of disaster, but once the disaster has occurred it is too late for effective planning. Communities and individuals can take steps **before** disasters occur to reduce the threat to lives and property.*

Most residents eventually made their way to shelters and were reunited with other family members and neighbors who could not be located during the evacuation. Others chose to stay temporarily with relatives or friends outside the disaster area.

After the disaster, many residents were eager to return home; but nothing, perhaps, could have prepared them for the conditions they had to confront on their return. In most of the affected homes, the water line on the walls was clearly visible, and a thick carpet of dirt and mud covered the floor. Most people found extensive damage to their personal belongings and homes. Some found that their insurance would not cover as much of their losses as they had assumed.

It became clear that the flooding had damaged parts of the community's streets and utilities, along with some public facilities and businesses. Small businesses, in particular, had suffered significant losses. Business owners were concerned about the future. Large amounts of inventory had been lost, and it would take a long time to repair structural damage. In the meantime, some businesses could not operate.

After the Centerville flood, residents returned to find extensive damage to their homes and property.

Federal, State, tribal, and local officials were working together, coordinating recovery efforts. A large-scale cleanup and recovery effort had begun even before the Presidential declaration. Debris removal teams were at work, and utility staff toiled to restore services. Businesses that were not severely damaged opened in an effort to get the community going and provide essential goods and services. But damage was severe and widespread; many families and individuals had to live in relatively poor conditions in spite of the relief provided. Electricity was not available in many areas for some time, and transportation out of the damaged areas continued to be difficult.

As they returned to their homes, residents were asking some basic questions:

◆ While my home is uninhabitable, where can I live?

◆ How can I get assistance to repair my home?

◆ How can my company recover from the loss of business as repairs are made?

◆ Where can I get help to restore my farm or ranch?

◆ What could I do to protect myself better from another disaster?

Announcements in all the media informed residents how to apply by telephone for assistance to repair damaged homes and businesses. Some were found eligible for financial assistance from the Federal and State governments, while others were able to rely only on insurance and personal savings.

As residents began to rebuild, some were already trying to think ahead. How could they be better prepared?

THINKING ABOUT DISASTERS

A Citizen's Guide to Disaster Assistance is intended to answer questions community members must face when disaster occurs. It will also suggest appropriate actions that could be taken prior to, during, and after a disaster to reduce injury and damage to property.

Unfortunately, some communities learn how to prepare for a disaster and reduce its effects only after having experienced one. It is then that they make major improvements in emergency plans, operational procedures, mitigation planning, and family preparedness activities and strengthen their defenses. But it is essential to begin planning *before* a disaster occurs. This course provides the benefit of insights drawn from the experiences of other citizens, communities, and disaster assistance professionals. It explains how planning and preparing *now* can lessen the effects of future disasters and the need for assistance.

Just what do we mean by an "emergency" or a "disaster?" For the purposes of this course, both terms refer to events that result in significant harm to lives and/or property, as well as disruption to normal patterns of living. Sometimes governments make important legal distinctions among the terms "emergency," "disaster," and "major disaster." Each government may define these terms differently for legal purposes.

Individuals and families can take actions that will reduce the injury to people and damage to property that often results from natural disasters (such as a flood) and technological disasters (such as a hazardous materials spill in a densely populated area). By making an effort to plan *before* an event (such as a flood) and identify the best actions to take *during* the occurrence, people can minimize the damages and disruptions and facilitate the return to normal *following* the emergency or disaster.

OVERVIEW OF DISASTER ASSISTANCE

Almost every one of us lives in a community that is vulnerable to some type of disaster, and many are exposed to several hazards. The threat of loss of lives and property is very real and national in scope. Government organizations at all levels—local, State, Federal and tribal—share the common goal of preventing or lessening the effects of disasters caused by earthquakes, floods, hurricanes, tornadoes, and other natural or technological events. Over the years, government agencies have worked with private and voluntary agencies to plan and coordinate disaster-related activities at all levels.

Individuals, families, and communities who are or may be affected by disasters are considered full participants in the preparedness process. Indeed, disaster assistance in the United States is provided within a framework that assumes every individual, family, and community will plan and respond within the limits of existing resources *before* other assistance may be made available.

This framework can be envisioned as a pyramid. Each step to higher authority is predicated upon the inability of the preceding authority to meet emergency needs adequately. Across this country, thousands of emergencies occur each year. The majority of these are handled by local government. Local emergency management crews—fire and police units, emergency medical and rescue personnel—provide immediate aid and protection to people and property.

Although three levels of involvement are possible in the disaster relief process, only unusually serious emergencies that require extensive resources result in requests for Federal—or even State—assistance.

Local emergency teams are joined by private and nonprofit organizations—the American Red Cross, the Salvation Army, churches, and other humanitarian groups—who provide emergency food, shelter, clothing, and other emergency needs. Public and private utility company crews move in to restore vital power, transportation, and communications lines.

If local officials need outside assistance, the Governor may find it appropriate to declare a state of emergency in the affected area, thus freeing state resources to provide the appropriate response and recovery activities. For example, the State National Guard can provide security, transportation, medical services, food, and temporary shelter. In Unit Two, you will learn more about how local and State governments, as well as voluntary agencies, help communities when emergencies occur.

At any time in a disaster, local government or State officials may turn to a number of Federal agencies for assistance. Many of these agencies can act quickly to provide some types of assistance under their own statutory authorities, independent of other Federal action. For example, the U.S. Coast Guard and the armed forces may assist in search and rescue operations if needed.

When a disaster situation is so severe that it is beyond the capabilities of local and State resources (even when supplemented by private and voluntary agencies and some limited Federal assistance), the Governor may request that the President declare a major disaster. Presidential approval of a Governor's request for Federal supplemental assistance activates a wide range of disaster assistance programs provided through several Federal agencies. Under the Federal Response Plan, representatives of these agencies will move into the disaster area to supplement response efforts if required. FEMA coordinates these response efforts and can also provide short-term and long-term recovery assistance.

While a wide range of Federal programs are available to aid disaster victims, it should not be assumed that all of them will be activated. The determination as to which programs will be provided is made based on actual needs found during the damage assessment and/or on the basis of subsequent information.

In Units Three and Four you will learn more about Federal assistance, which is the primary focus of this course. In Unit Three, "Overview of Federal Assistance," you will learn how the Federal government provides assistance to communities and the types of programs available. Unit Four, "Federal Disaster Assistance in Action," will describe how this help would be made available to you, your family, and your community.

Finally, in Unit Five, "The Citizen's Role in Disaster Preparedness," you will learn about steps you can take to prepare your family for disaster and reduce the risk of injury and financial loss.

COMMON MYTHS AND MISCONCEPTIONS ABOUT FEDERAL DISASTER ASSISTANCE

1. *The Federal government has total responsibility for disaster recovery.*

Local emergency response includes immediate aid and protection to safeguard lives and property and restore basic services, such as electric power.

The local government is primarily responsible for preparing for disasters that might affect a community and helping residents recover from such events. The great majority of disasters are handled successfully at the local level. State and Federal resources are intended to assist the community only when the community's own resources are not sufficient.

2. *The objective of Federal disaster assistance is to "fix everything."*

As much as we may wish otherwise, once a disaster has seriously impaired our homes and our communities, they may never be exactly the same. Nor will disaster assistance ever be adequate to restore everything that was lost by all those affected. The individual's own provision, especially insurance, must be used to ensure that losses can be recovered. Federal assistance will, in general, be used only for necessary expenses not met through other programs. Some of the Federal programs (such as loans from the Small Business Administration) cover most items that were lost, but not everyone is eligible. Other programs assist with only a portion of the losses or are intended only for serious needs.

3. *Everyone in the disaster area is eligible to receive Federal disaster assistance.*

   As part of the application process, applicants will have to demonstrate that they meet the eligibility requirements for each particular type of aid they are requesting.

4. *When Federal agency representatives arrive, they will immediately distribute money to disaster victims.*

   Individuals and families will need to plan to use their own resources and financial reserves until Federal funds can be released. An application process must be completed before assistance becomes available. Often, it takes several weeks for the Federal government to review requests for financial assistance and to issue funds to those who meet eligibility requirements. Most Federal assistance is in the form of a loan that must be repaid rather than an outright grant.

5. *Money received for Federal disaster assistance can be used as the recipient thinks best.*

   Monetary assistance is given for a specific purpose. The recipient must use the money to meet the need for which assistance was provided and must comply with specific regulations applicable to each type of assistance. If the assistance is in the form of a loan, recipients must be judged able to pay back the loan.

6. *Federal aid replaces the need for insurance.*

   Individuals, families, and businesses should all carry adequate insurance to meet their needs in the event of a disaster. It is not the purpose of Federal assistance to duplicate protection available through insurance plans. Federal assistance is provided to address only the most basic disaster-related needs not covered by other means. Besides, most disaster events are not Presidentially declared disasters, so Federal assistance is often not available.

Disaster assistance may be granted in the form of a loan rather than as an outright cash award. For geographical areas subject to floods, such as the one in the Centerville scenario, the Federal government ensures that residents of participating communities can receive appropriate insurance coverage through its National Flood Insurance Program (NFIP). In addition, flood insurance coverage is required as a condition to receiving Federal disaster aid for permanent repair or reconstruction of any structure located within an identified floodplain. You may wish to contact your local emergency preparedness office, the local building or zoning official, or your insurance agent to find out whether your local community is a participant.

Families that live in communities that participate in the National Flood Insurance Program are eligible to purchase flood insurance. This coverage is a condition to receiving Federal aid to permanently repair or reconstruct houses located in an identified floodplain.

SUMMARY

Systematic planning and action by local, State, Federal, and tribal governments are essential to ensuring effective response to, and recovery from, disasters. Reducing potential losses begins with hazard mitigation planning at the local level. Local officials are responsible for using resources appropriately to prepare for and deal with the emergency, while volunteer agencies supplement local resources in ways determined by the community plan. When local and volunteer agency resources are not adequate, State governments will assist local governments in dealing with the emergency. Likewise, voluntary agencies such as the American Red Cross may call on the resources of their national organizations. If the event is serious enough, the Federal government may provide additional assistance to supplement local and State resources.

In spite of many commonly held misconceptions, the Federal government will not assume total responsibility for fixing disaster damages, and everyone is not eligible for Federal assistance. Assistance that might be offered includes loans that would be available only to those with demonstrated needs and the ability to repay the loans. It is wise to carry appropriate insurance to ensure adequate reimbursement for losses. ◆

Check Your Memory
(Answers on page K-1)

1. Which of the following could be correctly described as the first line of defense in the event of an emergency?

    a. The Federal government.
    b. The State government.
    c. The local government.

2. The type of Federal disaster assistance provided depends largely on which of the following?

    a. Where the disaster occurs.
    b. The type of disaster that occurs.
    c. The duration of the disaster.
    d. Documented needs.

3. Most communities are vulnerable to some type of disaster.

    a. True
    b. False

4. Which of the following is a misconception about Federal disaster assistance?

    a. Assistance is generally in the form of loans.
    b. The objective of disaster assistance is to fix everything.
    c. The Federal government is by no means totally responsible for a community's disaster recovery.
    d. Everyone is not eligible for every form of Federal assistance.

5. The majority of emergencies that occur in the U.S. each year require Federal disaster assistance.

    a. True
    b. False

Unit Two

How Communities and States Deal with Emergencies and Disasters

During a flood such as Centerville's, many important activities must happen quickly and efficiently. Among these are rescue, caring for the injured, keeping people away from dangerous areas, assessing the situation to see what help is needed, and opening shelters for people displaced from their homes. As all this is occurring, phones are swamped with people asking for information. Without a good plan for such a situation, there would be no hope of getting the job done effectively.

In this unit, you will learn about:

◆ Local and State response, recovery, and mitigation activities.

◆ Local and State activities during the period following an emergency or disaster.

◆ The declaration of an emergency at the State and local level.

Fortunately, many local areas and States have developed emergency operations plans that help them respond and recover effectively. Their roles, as well as those of voluntary agencies whose invaluable efforts supplement theirs, are defined in these plans. When an emergency or disaster occurs, these plans are put into action to provide essential services to the community. The following information explains the types of activities performed by local government, State government, and voluntary agencies to deliver this assistance.

LOCAL RESPONSE AND RECOVERY ACTIVITIES

Local governments are the first line of defense against emergencies and disasters and are primarily responsible for managing the response to and recovery from those events. At the local government level, the primary responsibility for protecting citizens belongs to local elected officials such as mayors, city councils, and boards of commissioners. When a local government receives warning that an emergency could be imminent, its first priority is to alert and warn citizens and take whatever actions are needed to minimize damage and protect life and property. If necessary, it may order an evacuation. When an emergency or disaster does occur, fire and police units, emergency medical personnel, and rescue workers rush to damaged areas to provide aid. After this initial response, the local government must work to ensure public order and security. Vital services such as water, power, communications, transportation, shelter, and medical care must be provided, and debris removal must begin. Public and private utility company crews, along with other emergency teams, must be on the job to restore essential services. The local government coordinates its efforts with voluntary agencies who assist individuals and families in need.

Since disasters often disrupt water supply lines, local governments must ensure that residents receive drinking water.

When a local government responds to and recovers from a disaster, the levels of activities and the type of resources required are determined by several factors:

- ◆ The speed of onset of the emergency.
- ◆ The potential need for evacuation.
- ◆ The magnitude of the situation.
- ◆ The projected duration of the event.
- ◆ The extent of the threat to citizens.

Short-Term and Long-Term Recovery

In the aftermath of an emergency or disaster, many citizens will have specific needs that must be met before they can return to their pre-disaster lives. Typically, there will be a need for services such as these:

- ◆ Assessment of the extent and severity of damages to homes and other property.
- ◆ Restoration of services generally available in communities—water, food, and medical assistance.
- ◆ Repair of damaged homes and property.
- ◆ Professional counseling when the sudden changes resulting from the emergency have resulted in mental anguish and the inability to cope.

Local governments help individuals and families recover by ensuring that these services are available and by seeking additional resources if the community needs them. Also, when an emergency occurs, the local government uses all available media to publicize the types of assistance available and how to access them.

Recovery occurs in two phases—short-term and long-term. *Short-term recovery* measures are those that are intended to return the community to minimum operating standards. *Long-term recovery* are those steps taken to return to previous conditions (to the extent possible), combined with improvements that will better protect the community from future disasters. Each phase marks a transition that will enable the community to return to normal and create a safer condition for the future.

Following a disaster, repairing major roads is a high priority among short-term recovery tasks.

Short-term recovery could include making houses habitable so that families can move out of temporary shelters and return to their own homes. Short-term recovery also could involve restoring essential services so people can return to work. At the community level, this part of recovery may require completing repairs to roads and bridges so traffic can start moving again or restoring water and power to areas in need, especially to important public structures such as hospitals and major places of employment. In Centerville, for example, short-term recovery would include all of these activities, with priority given to restoring services in public structures and major places of employment. The restoration of major roads will be given priority to help people return to their homes and work safely; many minor routes may have to wait for repairs.

Long-term recovery may occur over a period of months or years, depending on the severity of the emergency or disaster. It often involves extensive repair and rebuilding. The disruption and destruction to the community can be so great that some businesses may never reopen or may have to relocate. Although a community may appear to be "open for business" a few weeks after an emergency or disaster, it may be years after a severe disaster before the community returns to pre-disaster conditions.

As part of the recovery, communities should consider strategies that would lessen the effects of a similar event in the future. These strategies, called mitigation measures, may have helped lessen the effects in the Centerville flood scenario. During the rebuilding process, residents could raise their furnaces to higher floors, business owners could consider storing inventory in areas above the flood level, and hospitals could elevate and move generators and other critical facilities to protected buildings. In the case of severe and repeated flood damage, residents might consider relocating damaged structures to a safer area. The community of Centerville could begin enforcing more stringent building codes and floodplain ordinances that help structures withstand flooding.

In addition to the self-help efforts of individuals and families and the efforts of local governments in emergencies, voluntary agencies are a central part of the effective response to, and recovery from, an emergency.

The Role of Voluntary Agencies

When most Americans think about disasters, they picture volunteers from agencies such as the American Red Cross and the Salvation Army providing a helping hand to the victims. Voluntary agencies are an essential part of any disaster relief effort, providing critical assistance with food, shelter, clothing, household items, medical expenses, clean-up, repairs, and rebuilding. These agencies are typically involved in all the phases of emergency management (mitigation, preparedness, response, and recovery).

Some voluntary agencies are available to assist in emergencies in all communities; others may assist only in disasters that affect specific regional areas. Voluntary agencies assist whether or not there is a Presidential declaration, coordinating with each other and with government officials to meet a community's disaster needs.

In a flood such as Centerville's, the State might be asked to help rescue stranded people and farm animals.

See Appendix B, page B-1, for a list of voluntary agencies that are active in disasters and the services they provide.

Requesting State Assistance

If the situation warrants, the community may have to reach beyond its own boundaries for additional resources. Mutual aid agreements should already be in place to facilitate provision of assistance by neighboring jurisdictions. In our flood scenario, however, these agreements will not result in significant added resources because other communities are also overwhelmed.

In such a case, the local government would have to appeal to the State for assistance. Centerville would seek assistance in transportation and rescue, for example. Local officials will submit a request to the Governor providing specific information about the situation and its effects and specifying the type of assistance needed. The State emergency management office and other offices involved in providing disaster assistance carefully assess this request and advise the Governor on appropriate actions.

Periodically, local officials send reports to the State that convey important information about the types and levels of assistance that might be required to assist the people in the impacted area. A typical situation report would contain information about the magnitude and severity of damages associated with the disaster event. Deaths, injuries, property damages, and locations in which losses occurred would be described. As additional information becomes available, updated reports are provided.

Generally, State emergency officials work very closely with local officials to ensure that required documentation is included in situation reports. If a request were to be made subsequently for a Presidential declaration (as will be explained in a later unit), the information contained in these reports would be of critical importance. The documentation of the local government's level of effort in responding to the event and the location of areas of damage are especially important.

STATE RESPONSE AND RECOVERY ACTIVITIES

All states have laws that describe the responsibilities of State government in emergencies and disasters. These laws provide governors and State agencies with the authority to plan for and carry out the necessary actions to respond to emergencies and recover from their effects. Typically, State emergency management legislation describes the duties and powers of the Governor, whose authority typically includes the power to declare a state of emergency and to decide when to terminate this declaration.

Many of the specific responsibilities to carry out the provisions of the State emergency management legislation are generally delegated to the State emergency management organization. Virtually all States have emergency management organizations, although their name and structure may vary from State to State. Typical names include office of emergency services or division of emergency management. Regardless of the title or location of the emergency management organization in the structure of the State government, its responsibilities are the same—to prepare for emergencies and to coordinate the activation and use of the resources controlled by the State government when they are needed to help local governments respond to, and recover from, emergencies and disasters.

The State emergency management organization, in its coordinating role, is involved in virtually all serious emergencies or disasters. Typically, this organization is responsible for receiving reports from the local area. Based on these and other data, emergency management officials work in consultation with other agency representatives and members of the

A Governor may declare a State emergency in order to facilitate the deployment of State resources to a disaster area.

Governor's staff to determine what types of resources and personnel should be deployed to the impacted area. Using procedures specified in the State plan, the State emergency management organization will coordinate deployment of State personnel and resources to the impacted areas.

However, it is not necessary for a Governor to declare an emergency or disaster before agency personnel and resources are deployed to monitor situations and provide information. Personnel and equipment are typically used to monitor situations in which an area's water supply may become contaminated or when large-scale chemical leakage is possible. State agency personnel would generally be involved in early inspection activities and in making reports back to the emergency management office and their own agencies for the purpose of determining additional assistance that may be needed.

When an emergency is declared, the Governor (or emergency management agency official acting for the Governor) can mobilize resources to supplement their own supplies, equipment, and personnel. In a situation like that of Centerville, for example, in which large populated areas are threatened by the continued rise in floodwaters, the State could assist in evacuation of the threatened area by prescribing evacuation routes and helping to control entries and departures from the disaster area.

State and local government also may regulate the movement of persons inside the affected area; persons can be prevented from returning to buildings rendered uninhabitable or unsafe by the disaster itself. The exercise of these powers could become necessary not only to protect the residents of the affected community but also to make the work of the emergency response personnel safer and more efficient.

In many States, governors can suspend State laws or local ordinances if it is determined that the law in question will restrict or prohibit efforts to relieve human suffering caused by the situation. In some States, after a State emergency declaration, the Governor may establish economic control over resources and services such as food, wages, clothing, and shelter in the affected area.

Under a State emergency declaration, Governors typically are empowered to mobilize the National Guard and direct its efforts.

Under a State emergency declaration, governors typically are empowered to mobilize the National Guard and direct its efforts. Generally, they are granted the power to use all available State resources needed to respond effectively and efficiently to the event. The Governor is able to draw upon the resources, expertise, and knowledge of State agencies as needed to assist in the effort. In many disasters, the States can provide technical assistance and resources that would not be available to most local officials within their own communities.

An affected State also is able to request mutual aid from other States. Participating States agree to provide personnel, equipment, and supplies to another State in need through the Emergency Management Assistance Compact (EMAC) or a similar arrangement. Mutual aid enables States to draw upon a common pool of resources with minimal Federal involvement.

Under a State emergency declaration, the Governor may also have the power to use or commandeer private property for the purpose of responding to the disaster. Emergency management acts generally grant the Governor the power to use, or authorize the use of, contingency and emergency funds in the event of an emergency. In some States, the Governor also may reallocate funds when designated funds are exhausted.

Types of Assistance Provided

Typically, there are two types of State response and recovery assistance.

◆　State personnel and resources can be activated and deployed to assist in the response and recovery effort directly (or to manage it, in some instances).

State officials may inspect structures such as dams, levees, and bridges to determine whether repairs are needed.

Examples of this type of activity include evacuation management, securing the affected area, and search and rescue.

◆　State personnel and equipment can be deployed to perform a variety of monitoring and inspection activities that can ensure the safety of inhabitants and response personnel in the area.

Examples of these types of activities include the use of officials to monitor threats of chemical and other fires or to monitor the water supply and ensure its continued safety. State officials may periodically inspect structures such as dams, levees, and bridges to monitor their condition and determine whether they are safe for continued use.

If necessary, the State may undertake emergency repairs (such as to restore bridges that are part of an essential route).

State assistance to communities is provided by many different State agencies. Some key agencies and types of services they provide are described below.

Department of Public Safety

State public safety officials may assist in search and rescue operations.

In many major floods—such as the one depicted in the scenario—bridges often are damaged, and very heavy debris may clog up the river, creating a more serious threat to surrounding areas. Heavy cranes and other equipment, along with the expertise and skill needed to use the equipment, can often be provided by State transportation or highway agencies. Engineers employed by transportation departments also have the knowledge and skills to conduct accurate damage assessments of bridges and other structures. In addition, they can suggest mitigation methods so that reconstruction includes added protection for future disasters.

State public safety personnel can assist in law enforcement for disaster areas, traffic control (especially in evacuation and for incoming assistance), security (such as to protect evacuated homes and businesses from looting and further damage), and search and rescue. The fire marshal's office can deploy personnel to investigate structural fires and to assist in assessing the safety of structures that may be at risk from fires.

Public health units within the public safety department often must perform tasks such as water supply monitoring, food supply inspection, and communicable disease control. State specialists also may assist in documenting (videotaping) damage.

Social Service Agencies

State social service agencies can provide or fund personnel and resources to assist in the management of shelters and to provide assistance to individuals and families. This can include counseling to alleviate stress, which, experience has demonstrated, must be handled appropriately in the early phases of a disaster to minimize later negative effects of the experience. If there is a Presidential declaration, these are usually the agencies that provide administrative services to manage the Individuals and Households Program. These agencies also are usually assigned to assist voluntary agencies such as the American Red Cross in their efforts to provide relief to disaster victims.

The National Guard

In a flood as serious as the one described in the scenario, the State National Guard could send personnel who could be assigned a wide range of duties. They would assist in flood-fighting activities such as sandbagging, evacuation, and search and rescue. The National Guard is frequently assigned to maintain order and civil control and to provide supplemental law enforcement and fire suppression assistance.

The National Guard units also have other valuable resources and equipment that can be used: trucks, helicopters, heavy tools and equipment, portable medical facilities, mobile kitchens, and communications equipment.

Public Health Agencies

State public health agencies perform several important functions in response and recovery. These agencies can make available: physicians, nurses, epidemiologists, medical technicians, and others. Equipment and facilities also are provided.

Monitoring water supplies, inspecting food supplies, controlling communicable diseases, providing and allocating medication in disaster-impacted areas, monitoring health care facilities, and identifying victims are among the more important response and short-term recovery activities that can be provided by, or coordinated through, State public health agencies.

State natural resource agencies may be able to contribute expertise when natural resources are threatened by fire.

Department of Agriculture

The State's department of agriculture will generally assist when damage to farms and ranches is involved. It often carries out measures to protect the long-term food supply of the affected area. State agriculture departments also inventory food resources and may help procure food for disaster victims. Longer term assistance provided by agriculture departments includes advising farmers and agribusinesses in mitigation planning and recovering from damages to facilities, crops, and livestock.

Natural Resource Agencies

Natural resource agencies have several types of expertise useful to an effective response, including fire suppression and the protection of fish and game resources. Natural resource agencies may have personnel available to assist in conducting damage assessments. Also, these agencies advise local officials and help them monitor and protect natural resources such as fish and game, as well as wildlands and other protected areas. Environmental protection agencies may assist in similar ways to help local officials preserve and protect various environmentally sensitive areas and to plan mitigation measures for further disasters.

They can also provide technical expertise to help agencies respond appropriately to hazardous materials spills that could result from primary events such as floods.

Other Resources

Other State agencies have resources and expertise helpful to local communities stricken by disaster. For example, labor departments can assist with immediate safety inspections. Education departments can help maintain education services. State management and budget agencies can assist in locating and establishing recovery centers and field operations offices.

Depending upon the severity of the disaster and the damages, some agencies— such as offices of management and budget, labor, employment security, commerce, and treasury—become more substantially involved in providing assistance for the community's recovery. For example, treasury departments can conduct post-emergency audits to document expenditures by local governments. In some States, they also provide tax advice for disaster victims.

Some State general services agencies can help identify and make available State facilities and related equipment to be used for shelter, as well as for the warehousing of food supplies or other resources.

In most States, commerce departments assist in licensing motor carriers and other vehicles needed to transport supplies. They also work to expedite and prioritize the recovery of utilities to the affected areas. Personnel from these agencies also may be involved in damage assessment work.

Finally, a key activity of State emergency offices is to review and critique the State's effort, with the objective of strengthening the State's response in the event of another disaster.

Requesting Federal Assistance

What if the available resources and personnel of both the local and State governments are inadequate to meet the response and recovery needs created by the disaster? The local government or State officials may at any time request assistance directly from a number of Federal agencies, most of which can provide some form of direct assistance without a Presidential declaration. When a disaster situation is beyond the capabilities of local and State resources, even as supplemented by private and voluntary agencies and by direct assistance from Federal agencies, the Governor may ask the President to declare a major disaster. If granted, supplemental disaster assistance is made available to help individuals, families, and the community.

THE ROLE OF PLANNING IN DISASTER ASSISTANCE

The ability of communities and States to effectively respond and recover from disasters depends largely on actions taken before the disaster. Communities and States should develop response, recovery, and hazard mitigation plans. Officials plan what roles different organizations would have in a disaster and how they would coordinate with each other to avoid duplication of benefits or confusion.

> **TERM TO REMEMBER**
>
> **Emergency Operations Plan**
> **or**
> **Emergency Response Plan**
>
> A document that contains information on the actions that may be taken by a governmental jurisdiction to protect people and property before, during, and after a disaster

Response Planning

The community's ability to respond to an emergency begins with the development of a local emergency operations plan.

Each community's plan may include a list of resources the community would use for various types of emergencies. In a flood such as Centerville's, for example, the local government will contact technical experts who can assess the condition of the flood protection structures and analyze the implications of their condition for flood control. The community's advance planning should also identify what routes could be used to evacuate people quickly in the event of a disaster. In Centerville's case, since the area has always been vulnerable to flooding, these would be pre-selected to facilitate movement. Shelter locations would also be identified as a preparedness measure.

The plan also establishes ways to notify the public in the event of an emergency. In our scenario, electric power was lost in Centerville, so many people could not get information either by phone or by television. The area did not have a siren warning system, and evacuation routes were not generally familiar to the public. As a result, the only means of reaching many people was through broadcasts that could be received only on battery-operated radios or by traveling to their neighborhoods.

Community plans should specify sources for the heavy equipment needed for debris removal.

Where did Centerville get the boats it used to rescue its citizens? Unless the town had worked out an advance agreement with possible sources, it lost valuable time trying to make the necessary arrangements at the time of immediate need. In addition to making arrangements with private sources to borrow resources (such as heavy equipment that may be needed for debris removal), communities may also have mutual aid agreements in place with adjacent communities to facilitate requests for assistance.

In addition to providing policies, procedures, and an emergency organization structure, the plan contains information on the specific emergency conditions under which the plan will be activated. If the conditions warrant, local authorities may declare an emergency. The legal basis for a local state-of-emergency declaration typically is a local ordinance that stipulates who has the authority to declare a state of emergency and under what conditions this can be done. Documentation provided in the plan gives local governments a solid legal foundation for any subsequent request for State and Federal emergency assistance and eliminates any confusion about the degree of impact the event has had on the community. Communities that formulate sound plans, establish appropriate emergency-related policies, and test their plans through regularly scheduled exercises will be prepared to assist citizens if an emergency occurs.

State governments also must document their plans for emergency response. The typical State plan is similar in structure and organization to most emergency operations plans developed by local governments. State and local plans should be coordinated to ensure that procedures for providing assistance result in an effective combined effort.

Recovery Planning

While State and local governments are experienced in developing and testing emergency response plans, only recently has the need for disaster recovery planning gained increased attention.

A disaster recovery plan establishes the roles, responsibilities, policies, and procedures to be used by State and local governments during the short- and long-term phases of a disaster. The disaster recovery plan may be separate from the emergency operations plan or it may be an annex to it. Some States require their local jurisdictions to develop disaster recovery plans or annexes and determine the issues to be included. Generally, however, the disaster recovery plan or annex should identify the roles and responsibilities of local government staff involved in disaster recovery operations, the organizational structure for the local disaster recovery staff, and policies and procedures that will be used during disaster recovery operations.

Examples of activities covered in disaster recovery plans are: debris removal, building inspection, public health and safety, temporary housing, temporary and permanent restoration of community services, disaster staffing, and documentation of expenditures for recovery operations.

State and Local Hazard Mitigation Planning

The Stafford Act requires that the recipients of disaster assistance make every effort to mitigate the natural hazards in the area. To comply with this provision, State and local governments must prepare and implement a hazard mitigation plan outlining cost-effective strategies to reduce vulnerability to specific hazards. Through the plan, State and local government can:

◆ Evaluate the hazards in the disaster area.

◆ Identify appropriate actions to mitigate vulnerability to these hazards.

The Stafford Act specifically encourages regulation of land use and protective construction standards as part of a long-term, comprehensive approach to mitigation. The President is also authorized to prescribe hazard mitigation standards and approve such standards proposed by State and local governments. Disaster assistance can be made conditional upon a recipient's agreement to develop a long-term strategy and program that will reduce or eliminate the need for future Federal disaster assistance should a similar event recur.

After a Presidential disaster declaration, FEMA works with the State to develop an Early Implementation Strategy. The strategy outlines activities to help reduce future damages based on damages assessed in the current disaster. This ensures that communities, States, and individuals consider ways to reduce potential damages from the next disaster as they make repairs now.

In the next unit, you will learn about Federal assistance and the conditions under which it is made available.

TRIBAL POLICY

Because of their unique status in the United States with the rights and benefits of sovereign nations, American Indian and Alaska Native Tribal governments have been assigned a separate disaster policy by FEMA that differs from that of the State governments. Once the President approves the State Governor's request for a disaster declaration, Tribal Governments that represent areas that sustained disaster damage can apply for disaster aid. Depending on the particular tribe and State, the application for disaster assistance will go either directly to FEMA or it will go through the State emergency management agency.

Disaster aid to Tribal Governments is authorized under the Stafford Act. This act authorizes FEMA to provide grants to individuals who do not qualify for other assistance for their unmet necessary expenses and serious needs. In addition, FEMA has a Public Assistance program that provides supplemental grant assistance to help Tribal Governments rebuild after the disaster. Finally, the Stafford Act also created the Hazard Mitigation Grant Program that provides grants to implement long-term hazard mitigation measures after a disaster declaration. These grants are provided on a cost shared basis and normally the recipient provides 25% of the cost of the project.

SUMMARY

Local governments are the first line of defense against emergencies. When needed, they serve as the link between individuals and the emergency response and recovery efforts carried out by State and Federal government.

Response involves immediate actions to save lives, protect property, and meet basic human needs. Short-term recovery generally involves temporary measures to restore essential services and get the community going again. Long-term recovery involves permanent restoration, including steps to provide greater safety for the future. Local ordinances and emergency operations plans are the basis for the local response effort. Voluntary agencies are an integral part of the community response effort.

The local government requests State assistance when it is needed. The State uses local reports describing damages incurred and local actions taken to determine how to best direct its resources.

If the State's resources are also overwhelmed, the Governor may request specific types of assistance from the Federal government. ◆

Check Your Memory

(Answers on page K-2)

1. To protect citizens in an emergency, local government _____.

    a. Activates its local emergency operations plan.
    b. Warns citizens.
    c. Reports to State officials.
    d. All of the above.

2. Large-scale restoration and replacement of buildings or roads are _____ activities.

    a. Short-term recovery.
    b. Long-term recovery.

3. Situation reports are used to _____.

    a. Monitor local response.
    b. Evaluate the extent of damage.
    c. Identify needed assistance.
    d. All of the above.

4. During an emergency or disaster, the Governor may be authorized to _____.

    a. Activate Federal assistance.
    b. Mobilize State agency resources.
    c. Appropriate private resources.
    d. Both b and c.

5. The State National Guard could assist communities in _____.

    a. Food supply inspection.
    b. Maintaining order.
    c. Counseling.
    d. Advising farmers on damage recovery.

6. After a disaster happens, there is no point doing anything to reduce the damages that might occur next time.

    a. True
    b. False

Unit Three

Overview of Federal

Disaster Assistance

Federal assistance is available to supplement the resources of State, local, and voluntary agencies in major disasters. Some forms of Federal assistance could be available without a Presidential declaration. Others would become available only following a declaration by the President at the request of the State's Governor. FEMA uses the Federal Response Plan (FRP) to coordinate the government response to disasters or emergencies. The FRP describes the mechanisms by which the

---

In this unit, you will learn about:

◆ The role of the Federal government in disaster assistance

◆ Types of Federal assistance available in disasters through the Federal Response Plan (FRP).

◆ The disaster response and recovery cycle.

◆ Eligibility criteria for major types of assistance.

◆ The purpose and function of an Emergency Response Team.

Federal government mobilizes resources and conducts activities to augment State and local response efforts.

THE ROLE OF THE FEDERAL GOVERNMENT IN DISASTER ASSISTANCE

To understand the role of the Federal government in disaster relief, it is worthwhile to briefly review the history of its involvement.

During the period from 1803 to 1950, Congress passed 128 separate laws dealing with disaster relief. Because there was no comprehensive legislation covering disaster relief, Congress had to pass a separate law to provide Federal funds for each major disaster that occurred. The system was a cumbersome one.

In 1950, Congress passed the Federal Disaster Relief Act (Public Law 81-875), authorizing the President to provide supplementary Federal assistance when a Governor requested help and the President approved the request by declaring a major disaster.

A critical statement in the act established the philosophy of the nation's disaster response and recovery program. Federal disaster assistance would "supplement the efforts and available resources of the State and local governments." In other words, the act made it clear that the Federal government would not function as the first-line provider of emergency assistance and disaster response and recovery. It would *support* State and local governments—not *supplant* them. To further underline this philosophy, the act required that Federal assistance be supplied when, and only when, State and local governments had themselves committed "a reasonable amount of the funds" needed.

In 1968, the Federal government took another step aimed at benefiting communities; but again, it was linked to steps that had to be taken by communities. This was the year that the National Flood Insurance Act was signed into law. Community participation requires adoption and enforcement of prudent, flood-resistant construction techniques for all new, substantially improved, and substantially damaged structures located within identified floodplain areas. The Act gave individuals and communities a way to reduce their reliance on the Federal government and take personal responsibility for their own recovery.

When Hurricane Agnes swept through the eastern part of the United States in 1972, it caused unprecedented levels of damage. The effects of this hurricane led the Federal government to reexamine existing legislation and address weaknesses, particularly in the area of assistance to individuals. The hurricane was the most costly natural disaster that had occurred to date in the country. It caused disastrous floods and flash floods almost simultaneously over the eastern seaboard—a feat unique in the country's abundant experience of natural disasters. One result of the disaster was that it motivated Congress to strengthen certain provisions of the National Flood Insurance Act. Among other changes, the Act was revised to require Federally insured lending institutions to require flood insurance on new loans for homes and other property in designated floodplains (areas vulnerable to flooding).

A second major disaster in 1974 again spurred the Federal government to action. On "Terrible Tuesday"—April 3—tornadoes struck across 10 states, resulting in six Federal disaster declarations. As a result, the Federal government passed the Disaster Relief Act of 1974, which consolidated many changes that had been initiated in the period following Hurricane Agnes.

During President Carter's administration (1976-1980), the Federal government undertook an extensive evaluation of its disaster response and recovery programs to determine which of them could be combined to increase efficiency and save money. The Federal Emergency Management Agency (FEMA) was created in 1979, combining under its roof a number of emergency management programs that had been administered by different agencies.

The first disaster to be funded on a cost-sharing basis—75 percent Federal and 25 percent non-Federal—was the 1980 eruption of Mount St. Helens, which deposited a blanket of ash throughout Washington State and in other parts of the West. In 1988, the Robert T. Stafford Disaster Relief and Emergency Assistance Act (Public Law 93-288, as amended) legislated cost-sharing requirements for public assistance programs. It also provided funds for states and local governments to manage public assistance projects, encouraged hazard mitigation through a new grant program, and gave the Federal government the authority to provide assistance for disasters regardless of cause. Cost-sharing requirements continue to be a cornerstone of Federal disaster assistance policy.

AUTHORITY FOR DISASTER ASSISTANCE

Today, the Robert T. Stafford Act gives the Federal government its authority to provide response and recovery assistance in a major disaster. The Stafford Act identifies and defines the types of occurrences and conditions under which disaster assistance may be provided. Under the law, the declaration process remains a flexible tool for providing relief where it is needed.

The Federal Response Plan (FRP), created in 1992, describes how the Federal government will mobilize Federal resources and conduct activities to assist State and local governments in responding to disasters. The Plan relies on the personnel, equipment, and technical expertise of 27 Federal agencies and departments, including the American Red Cross, in the delivery of supplemental assistance. FEMA is responsible for the Plan's overall coordination.

The Department of Homeland Security (DHS) Act of 2002 (Public Law 107-296) transferred FEMA and a number of other Executive Branch components and functions to the new Department. Subsequently, the Interim FRP dated January, 2003, was published to reflect the provisions of the law impacting the FRP.

Of course, the State does not always request Federal assistance. A great many disasters are handled successfully at the State and local levels with the assistance of voluntary agencies and private agencies. Although the exact number of disasters successfully handled without requests for Federal assistance is not known, it is estimated at 3,500 to 3,700 annually.

A Presidential disaster declaration is the result of a legal process involving specific steps and actions taken by local, State, and Federal governments. These steps are depicted in the graphic on page 3-8, which shows an overview of the declaration process.

In the flood scenario presented at the beginning of this course, local officials declared a State of Emergency, acting in accordance with the local emergency operations plan. As the flood waters rose and spread and essential buildings suffered major damage, local officials determined that they did not have adequate resources to respond effectively to a flood of this size and asked the State for assistance. To support their request, local officials described the extent and types of damage caused by the flood. They asked for specific kinds of assistance, including help in evacuating persons from affected areas and in keeping people from entering unsafe highways or other restricted areas.

Typically, when a disaster as serious as that in the scenario occurs, it is apparent from an early stage that not only State but also Federal assistance may be needed. State and FEMA officials would continually monitor the progress of the incident. Under the FRP, the Federal governments is prepared to provide supplemental assistance to State and local government in 12 major areas known as emergency support functions, or ESFs. Each ESF is assigned to a primary agency, supported by as many as 17 other support agencies with similar missions and responsibilities. These include the following:

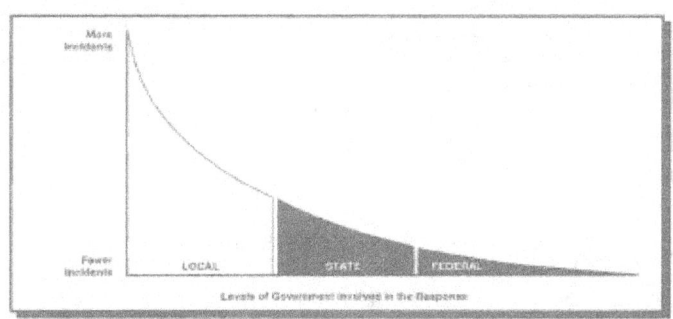

Most emergencies are handled at the local level, without assistance from the State or the Federal government. Only a small number result in a request for Federal assistance.

ESF# 1   Transportation

2   Communications

3   Public Works and Engineering

4   Firefighting

5   Information and Planning

6   Mass Care

7   Resource Support

8   Health and Medical Services

9   Urban Search and Rescue

10   Hazardous Materials

11   Food

12   Energy

```
TERM TO REMEMBER
Emergency Support Function
(ESF)

One of the 12 areas in which the
Federal government is organized
to provide support to State and
local responders in a disaster.
```

In some instances, Federal personnel representing some or all the ESFs may be activated even *before* a disaster occurs. The advance element of the Emergency Response Team, called the ERT-A, is headed by a FEMA team leader and is composed of FEMA program support staff and ESF representatives. In predictable disasters, such as hurricanes, ERT-A teams may be sent into the area before the storm strikes to set up emergency communications equipment and help coordinate early response efforts. An important role of the ERT-A is to obtain information on the impact of the event and identify the types of short and long-term assistance that may be needed.

When a major disaster occurs (or before, for predictable events), the Regional Support Team (RST) staff is activated to the ROC by the Regional Director of the FEMA regional office. The RST staff is the initial coordination organization for Federal activity. Other elements of the initial Federal response may include the Emergency Support Team (EST) and the Emergency Response Team (ERT).

The ERT is the Federal interagency team which is activated following a Presidential disaster or emergency declaration. The ERT coordinates the overall Federal response and recovery activities and provides assistance and support to the affected State and local governments. The ERT is headed by a Federal Coordinating Officer (FCO) and operates from a Disaster Field Office (DFO). The size and composition of the ERT can range from one that includes only FEMA regional staff, to an interagency team with representation from ESF primary and support agencies.

The EST is an interagency group that operates from the National Interagency Emergency Operations Center (NIEOC) located at FEMA Headquarters. Its role is to coordinate and support the Federal response by serving as an information source and by helping access and coordinate needed resources.

Meanwhile, local and State responders are fully committed as they attempt to respond to a major disaster. Local first responders work closely with voluntary agencies; the Mayor or County Executive activates the local EOC. Upon a request from the local executives, the

Governor activates the State EOC, declares a State emergency or disaster, and activates the State Emergency Operations Plan.

If early damage reports lead the State to conclude that effective response may exceed both the State's resources and those of the community, the State can request that FEMA regional officials join them in conducting joint *preliminary damage assessments*—known as PDAs—in areas designated by the State officials. FEMA has 10 regional offices, each responsible for specified states. Appendix A (page A-1) shows the FEMA regional structure.

> **TERM TO REMEMBER**
>
> **Preliminary Damage Assessment (PDA)**
>
> The joint local, State, and Federal analysis of damage that has occurred as a result of an incident, and that may result in a Presidential declaration of disaster. The PDA is documented through surveys, photographs, and other written information.

The data gathered in these joint assessments are used for several important purposes in the Presidential disaster declaration process:

◆ Determine the impact and magnitude of damage incurred.

◆ Determine resulting unmet needs of individuals, families, and businesses, as well as the impact to public property.

◆ Document that the disaster is beyond local and State capabilities and support the Governor's request for Federal assistance.

◆ Provide the basis for FEMA's recommendation to the President.

◆ Determine the types of assistance needed and the areas where assistance should be offered.

◆ Determine the extent of the Federal government's commitment (including staff, equipment, and funding).

◆ Provide essential management information to State and Federal disaster officials.

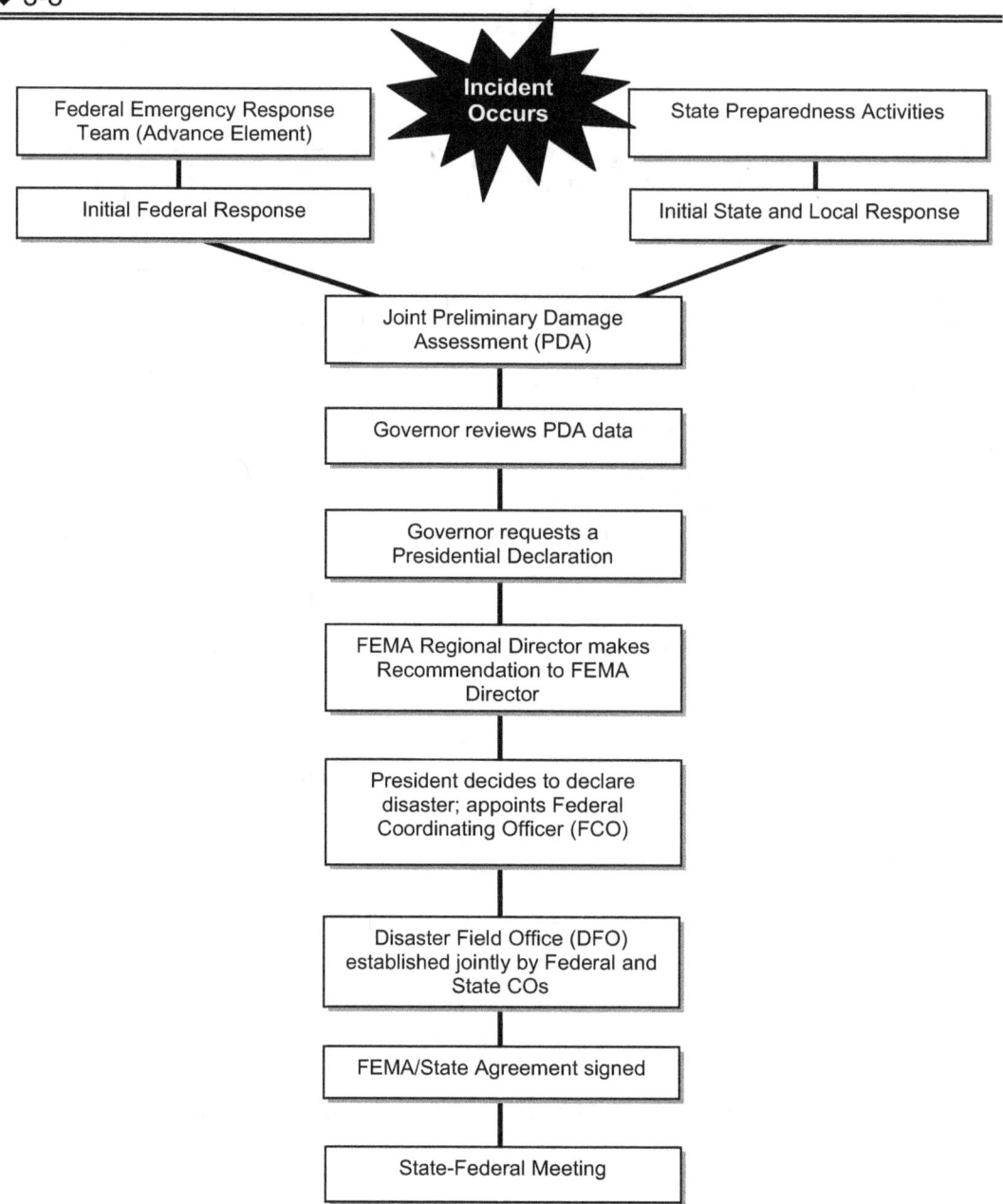

Federal Emergency Response Team (Advance Element)

Incident Occurs

State Preparedness Activities

Initial Federal Response

Initial State and Local Response

Joint Preliminary Damage Assessment (PDA)

Governor reviews PDA data

Governor requests a Presidential Declaration

FEMA Regional Director makes Recommendation to FEMA Director

President decides to declare disaster; appoints Federal Coordinating Officer (FCO)

Disaster Field Office (DFO) established jointly by Federal and State COs

FEMA/State Agreement signed

State-Federal Meeting

The Presidential disaster declaration process begins with a request from the Governor of the affected State; the response is ultimately determined by the President.

The amount of information collected may vary. In general, the larger and more severe the disaster, the less information is needed initially to support the request. A smaller or less obvious situation requires a greater amount of information to verify that Federal assistance is needed. An aerial survey conducted by FEMA and State officials might be enough to demonstrate the need for Federal help, although damage assessments would continue after the declaration to help manage response and recovery measures.

After the PDA teams have documented the damage that has occurred and assessed its impact on the community, the Governor will determine whether or not to request Federal disaster assistance. In order to make the request, the State must have implemented its State Emergency Operations Plan. The request must include specific information required by law, including the nature and amount of local and State resources that have been or will be committed to disaster-related work. The request must also guarantee that the cost-sharing provisions of the Stafford Act will be met. FEMA regulations generally require that the request be submitted within 30 days of the incident, but a waiver may be granted at the Governor's request provided it is made within the 30-day period.

PDA data forms the basis for immediate funding for emergency work under the Public Assistance Program in communities hit hardest by a disaster. This *immediate needs funding*—up to 50% of the Federal share of PDA estimates for emergency work—provides funds for applicants to continue emergency protective measures and debris removal without the burden of extensive documentation and review during the peak of crisis operations.

Basic disaster assistance from the Federal government falls into three categories: assistance for individuals and businesses, public assistance, and hazard mitigation assistance.

A Governor's request may seek any or all of these. However, hazard mitigation assistance is implemented only if one of the other categories is designated available. Mitigation assistance can be used throughout the affected State, though, rather than just in the declared counties.

◆ *Assistance for individuals and businesses* includes assistance available to individuals, families, and businesses; it can include disaster housing, unemployment assistance, individual and family grants, legal services, crisis counseling, tax relief, and agricultural assistance. Small businesses may apply for low-interest loans for repairs.

◆ *Public Assistance* refers to programs that provide funding assistance and technical expertise to aid State and local governments and certain facilities of private, nonprofit organizations. Primarily, Public Assistance refers to funds for repairing or replacing essential public systems and facilities.

◆ *Hazard mitigation assistance* provides Federal aid in support of measures that will permanently eliminate or reduce an area's long-term vulnerability to the loss of human life and property from a particular hazard.

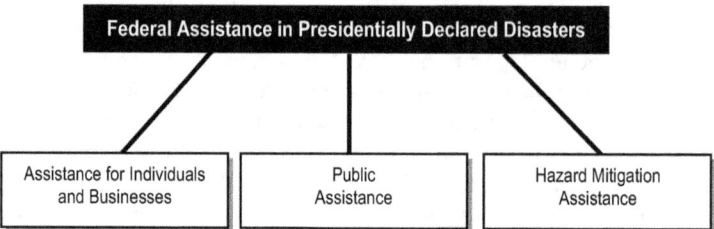

Basic disaster assistance from the Federal government falls into three major categories

A Governor's request for Federal aid is addressed to the President and forwarded to the appropriate FEMA Regional Director. This person evaluates the on-scene damage assessment information and the requirements for supplemental assistance and forwards a recommendation to the FEMA Director in Washington, D.C. for review. The FEMA Director's recommendation is then forwarded to the President.

In developing a recommendation, FEMA considers several factors:

◆ The amount and type of damage.

◆ The impact of losses on individuals, businesses, local governments, and the State.

◆ Available resources of State and local governments and voluntary agencies.

◆ The extent and type of insurance in effect to cover losses.

◆ Assistance available from other Federal agencies without a Presidential declaration.

◆ Imminent threats to life and safety.

◆ The recent disaster history in the State.

◆ Other factors pertinent to a particular incident.

WHEN A DISASTER IS DECLARED

When the President determines that a State requires supplemental Federal assistance, a formal disaster declaration is made. The Governor, members of Congress from the affected area, affected Federal departments and agencies, and the media all receive notice of the declaration.

After the President declares a major disaster, FEMA prepares a document called the FEMA-State Agreement. This agreement describes the period of the incident (or disaster), the types of assistance to be provided, the areas eligible for assistance, the agreed-upon cost-share provisions, and other terms and conditions.

Both FEMA and the State commit to the terms and conditions of the document. It may be amended if the situation changes; for example, additional counties may be included in the declaration, or an end date for the disaster may be specified.

TERM TO REMEMBER
FEMA-State Agreement

A formal, legal document between FEMA and the affected State that describes the understandings, commitments, and binding conditions for assistance applicable as the result of the major disaster or emergency declared by the President. It is signed by the FEMA Regional Director and the Governor.

FEMA's Role in Managing Disasters

Once a major disaster declaration has been made, the FEMA Director, under the authority of the President, will appoint a Federal Coordinating Officer (FCO). The FCO is responsible for coordinating the overall disaster response and recovery effort—including local, State, and Federal resources—to ensure that an adequate quality and quantity of disaster assistance is provided. The FCO also works in cooperation with voluntary agencies (such as the American Red Cross or Salvation Army) to avoid duplication of effort and ensure priority needs are efficiently met. Throughout the Federal disaster response and recovery operation, the FCO serves as the President's representative. The FCO's immediate concern after a major disaster declaration is to make an initial appraisal of the types of relief most urgently needed. The FCO coordinates all Federal disaster assistance programs to ensure maximum effectiveness, and takes appropriate action to help the community obtain the assistance needed.

A Disaster Field Office (DFO) is established in the disaster area for State and Federal staff. The office serves as the focal point for directing and coordinating the many different types of disaster operations underway and for maintaining the network among the many participating agencies. Here, the Emergency Response Team (ERT) is stationed to carry out the Federal role in providing the disaster assistance required. The DFO is *not* intended to receive the public; separate facilities are established for this function.

> **TERM TO REMEMBER**
> **Federal Coordinating Officer (FCO)**
>
> The person appointed by the FEMA Director (by delegation of authority from the President) to coordinate a Presidentially declared disaster.

> **TERM TO REMEMBER**
> **Disaster Field Office (DFO)**
>
> The office where Federal (and other State) disaster relief and recovery efforts are coordinated. It is staffed by the Emergency Response Team (ERT) composed of representatives of responding agencies.

At the State level, the State Coordinating Officer (SCO) has roles and responsibilities similar to those of the FCO. The SCO coordinates State and local assistance efforts with those at the Federal level. The SCO is the principal point of contact for State and local response and recovery activities and oversees implementation of the State emergency plan.

TERM TO REMEMBER
State Coordinating Officer
(SCO)

The individual appointed by the Governor to act in cooperation with the Federal Coordinating Officer (FCO) to administer disaster response and recovery efforts.

The Federal, State, private, and voluntary agency response team relationship is established and fostered at an initial meeting held as soon as possible after the President's declaration. All Federal, State, and voluntary agencies that can provide some form of disaster assistance are invited to be represented at this meeting. Initial relief coordination procedures are established, especially the details of setting up and staffing sites where disaster victims can apply for assistance. The FCO and SCO conduct subsequent coordination meetings as often as needed to establish priorities and objectives, identify problems, and document achievements.

TYPES OF FEDERAL ASSISTANCE AVAILABLE IN DISASTERS

The following section reviews some of the types of Federal disaster assistance that may be available in each of three categories: assistance for individuals and businesses, public assistance, and hazard mitigation assistance. General procedures for accessing this assistance will be discussed in the next unit.

Assistance for Individuals and Businesses

In many disasters, individuals, families, and small businesses suffer the most devastating damage. The following Federal programs could be made available to assist them.

Providing Food to Affected Individuals and Families

The Food and Nutrition Service is an agency within the U.S. Department of Agriculture (USDA) that oversees two major disaster assistance programs: food coupons and food commodities. Assistance in purchasing food is available through the Disaster Food Stamp Program. After national-level approval by the Food and Nutrition Service of the USDA, the responsible State and/or local social services agency would provide disaster food stamps to eligible households, who would apply through the local social services office.

Many volunteer agencies are able to help prepare and distribute food to disaster victims. The Secretary of Agriculture can assist by donating surplus commodities.

The Secretary of Agriculture has the authority to donate surplus commodities for the mass feeding of disaster victims. Eligible applicants are public or voluntary agencies or households on Indian reservations served by the Needy Families program. Individuals may receive food assistance through the American Red Cross, Salvation Army, Second Harvest Food Bank Network, Southern Baptists, and other organizations providing mass care.

Disaster Housing Assistance

The Federal government can make funds available to restore homes to a safe, sanitary, and functional condition. Homeowners must prove they owned and occupied the home at the time of the damage and that damage was disaster-related. The Disaster Housing Program can provide funds to be used in renting a place to live. Renters must prove that they lived in the disaster damaged house. Homeowners or renters who can prove they suffered financial hardship as a result of the disaster and cannot pay their rent or mortgage may also qualify for financial help to make those payments.

Business may be eligible for physical loss disaster loans intended to repair disaster-related damage to property—Including inventory and supplies—owned by the business.

Disaster Loans for Individuals and Businesses

Disaster victims whose property is damaged or destroyed by a disaster may be able to receive a loan from the Federal government to help with repairs. Even without a Presidential declaration of disaster, the Small Business Administration (SBA) may provide disaster assistance in the form of low interest loans to qualified individuals and businesses. To receive an SBA loan, applicants must demonstrate their ability to repay the loan. Disaster loans may be made available to homeowners to repair or replace homes or personal property. Renters also may be eligible for loans to repair or replace personal property damaged by the disaster.

Businesses may receive SBA *physical loss* disaster loans intended to repair disaster-related damage to property owned by the businesses, including inventory and supplies. *Economic injury* disaster loans provide working capital to small businesses and to small agriculture cooperatives to assist them through the disaster recovery period. These are available only if the business or its owners cannot obtain this type of assistance from non-government sources.

Homeowners who live in areas devastated by disaster may be eligible for special loans from the Farm Service Agency (FSA), an agency of the U.S. Department of Agriculture. These loans can be used to restore or replace essential property, and are available to established family farm operators who are unable to receive credit from commercial sources. Real estate (homes) may be eligible. Those seeking additional information or wishing to apply for assistance should contact their local FSA county office.

Individuals and Households Program

The two types of assistance available under the Individuals and Households Program are:

◆ Housing Assistance (HA)

◆ Other Needs Assistance (ONA)

Funds from the Individuals and Households Program could be used to remove debris that threatens to harm a residence.

Housing Assistance

The Federal government can make funds available to homeowners and renters for one or more of the following types of housing assistance:

◆ Rental Assistance

◆ Repair Assistance

◆ Replacement Assistance

◆ Permanent Housing Construction

Homeowners must prove they owned and occupied the home at the time of the disaster and that the damage was disaster related. Renters must prove that they lived in the disaster-damaged house at the time of the disaster.

Housing Assistance provided under IHP can be financial or direct. Financial Assistance is provided through grant funds and the amount varies with the type of assistance provided:

◆ Repair Assistance—up to $5,000.00 (CPI adjusted annually)

◆ Replacement Assistance—up to $10,000.00 (CPI adjusted annually)

◆ Permanent Housing Construction—mobile home or travel trailer. Assistance provided directly has no dollar limit, financial assistance is limited to the program maximum of $25,000.00 (CPI adjusted annually)

◆ Rental Assistance—up to the program limit of $25,000.00 (CPI adjusted annually) or 18 months of assistance, whichever comes first.

Housing assistance under IHP is not SBA dependent.

Other Needs Assistance

Other Needs Assistance is a FEMA/State cooperative venture that assists disaster victims with disaster-related serious needs and necessary expenses who have no other source of government, private, or insurance assistance available. Assistance is provided for various personal property losses when the applicant has been denied by SBA for a disaster assistance loan or provided a loan insufficient to cover the disaster-related losses. Covered items are:

◆ Household items, furnishings, and appliances.

◆ Clothing.

◆ Tools or specialized clothing and equipment required by an employer.

◆ Moving and storage of personal items to prevent further damage.

◆ Privately owned vehicles.

◆ Flood insurance coverage for a 3-year period.

Assistance that may be available under the ONA provisions that are not SBA dependent is:

◆ Public transportation or other transportation needs.

◆ Medical, dental, and funeral expenses.

◆ Other eligible miscellaneous expenses (e.g. generator, wet/dry vac, etc.)

Assistance for Farmers and Ranchers

Agencies of the U.S. Department of Agriculture can give assistance to farmers and ranchers even without a major disaster declaration by the President. The Farm Service Agency (FSA) Emergency Conservation Program (ECP) helps fund repair of fencing, debris removal, or restoration of damaged land by grading and shaping. During a drought, ECP also provides emergency water assistance, both for livestock and for existing irrigation systems for orchards and vineyards.

The FSA can provide financial assistance to eligible producers affected by natural disasters. The Noninsured Crop Disaster Assistance Program (NAP) covers noninsurable crop losses and planning prevented by disasters.

The FSA also can make emergency management (EM) loans in counties (or parishes) included in a Presidential disaster declaration or by the Secretary of Agriculture as a disaster area or quarantine area.

The Farm Service Agency can provide emergency loans when a natural disaster severely impacts a farming, ranching, or aquaculture operation.

EM loans may be made to farmers and ranchers who:

◆ own or operate land in a designated disaster area

◆ are established family farm operators with sufficient farming and ranching experience

◆ are citizens or permanent residents of the U.S.

◆ have suffered at least a 30% loss in crop production or a physical loss to livestock and livestock products, real estate, or chattel property

◆ have an acceptable credit history

◆ are unable to receive credit from commercial sources

◆ can provide collateral to secure the loan, and

◆ have repayment ability.

Disaster Unemployment Assistance

In a disaster such as the flood in Centerville, many businesses may temporarily cease to operate, and unemployment may be high. The Disaster Unemployment Assistance (DUA) program provides unemployment benefits and reemployment services to individuals who have become unemployed because of major disasters and who are not eligible for other unemployment compensation programs. The Department of Labor is authorized to administer the program, for which FEMA is responsible, under the Stafford Act. All unemployed individuals must register with the State's employment services office before they can receive DUA benefits.

Internal Revenue Service (IRS)

The IRS provides counseling on how to prepare or amend returns to include casualty loss deductions. Certain casualty losses may be deducted on Federal income tax returns through an immediate amendment to the previous year's return.

Legal Services

Low-income individuals who need legal assistance due to a disaster may be eligible for free legal consultation and services. This type of assistance may be provided by the Young Lawyers Division of the American Bar Association, the State Bar Association, or the State's Attorney, and coordinated through the FEMA Regional Director or the Federal Coordinating Officer. Assistance may include: insurance claims, lost legal documents, powers of attorney, and home repair contracts.

Social Security Benefits

The Social Security Administration (SSA) does not offer special disaster benefits. However, in a disaster, it is important that those who depend on Social Security checks continue to receive them, even though they may be displaced from their homes. Therefore, the SSA Regional Commissioner provides support staff to do the following:

◆ Process Social Security claims.

◆ Provide advice and assistance in regard to regular and survivor benefits payable through Social Security programs.

◆ Process disaster-related death certificates.

◆ Resolve problems involving lost/destroyed Social Security checks.

◆ Make address changes.

◆ Replace Social Security cards.

Assistance to Veterans

The Department of Veteran Affairs (VA) provides a variety of disaster assistance specifically targeted to veterans and survivors, including the following:

◆ Medical assistance.

◆ Burial assistance.

◆ Priority in acquiring VA-owned properties if you are displaced in a disaster.

◆ Health care supplies and equipment, drugs, medicine, and other medical items.

◆ Temporary use of housing units owned by the VA.

Crisis Counseling

In a major disaster, many people become stressed, grief-stricken, or disoriented. Imagine the grief of those recovering from a flood such as Centerville's as they return to devastated homes. The Stafford Act authorizes the President to provide funding for training and services to alleviate mental health problems caused or exacerbated by major disasters. The training is designed to supplement the available State and local government resources. There are two types of grants: immediate services funding and regular program funding.

◆ *Immediate services program* provides screening, diagnostic, and counseling techniques, as well as outreach services such as public information and community networking, to help meet mental health needs immediately following a disaster up to 60 days from date of the declaration.

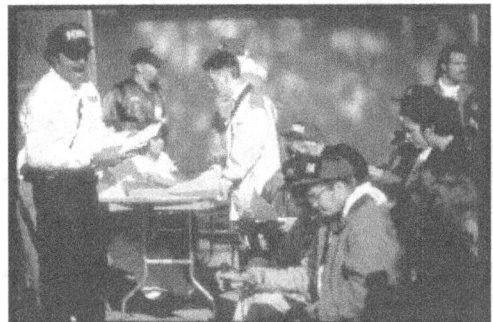

Federal grants may be used to provide funding for community outreach and education following a disaster.

◆ *Regular services program* provides funding for crisis counseling, community outreach, and consultation and education services to assist people affected by the disaster up to 9 months from the date of the declaration. These grants are provided by FEMA through the Center for Mental Health Services, part of the Department of Health and Human Services (HHS); they are usually administered by State health departments.

Cora Brown Fund

This fund is used to assist victims/survivors in Presidentially declared disasters with disaster-related needs not met elsewhere. Potential recipients do not need to apply for this assistance; rather, they are identified by FEMA representatives with assistance from other Federal, State, local, and voluntary relief agencies. Assistance that can be provided by the Cora Brown Fund includes: disaster-related home repair and rebuilding, health and safety measures, assistance to self-employed persons (with no employees) to reestablish their businesses, and other services which alleviate human suffering and promote well being of disaster victims.

Public Assistance

The preceding section describes the major types of assistance that could be made available to eligible *persons, families, and private businesses* under the declaration of a major disaster. This section provides information on the major types of assistance designed to meet *community* needs.

Federal funds may be used to repair or replace disaster-damaged public facilities, such as schools.

At the discretion of the President, FEMA can provide supplemental financial assistance and technical assistance to State and local governments, Indian tribes and certain private nonprofit organizations to help communities recover from disasters as quickly as possible. Through the Public Assistance Program, FEMA provides grants to eligible applicants for emergency protective measures, repair, replacement or restoration of damaged facilities not met by insurance. Eligible applicants include: State, local governments, Indian tribes and private non-profit (PNP) organizations that provide essential services of a governmental nature, such as medical facilities, emergency facilities, utilities, educational facilities, and custodial care facilities. Eligible work is classified as either "emergency work" or "permanent work."

Emergency Work

Emergency work includes those activities undertaken by a community before, during and after a disaster to save lives, protect public health and safety, and prevent damage to improved public and private property.

Examples of emergency work include, but are not limited to:

◆ Debris removal.

◆ Emergency protective measures to protect lives or improve property.

◆ Search and rescue.

◆ Demolition and removal of damaged public and private buildings and structures that pose an immediate threat to the safety of the general public.

Permanent Work

Permanent work refers to any activity that must be performed to restore a disaster-damaged facility to pre-disaster design, function and capacity. Examples of eligible facilities include:

◆ Roads, bridges, and associated facilities (except those on the Federal aid system).

◆ Water control facilities such as dams and reservoirs, levees, pumping facilities, and drainage channels.

◆ Buildings such as police stations, hospitals, schools, and libraries. Contents of buildings such as furnishings and interior systems, as well as equipment such as vehicles are also included.

◆ Utilities include systems for water treatment and conveyance, power generation and distribution, sewage collection and treatment, and telecommunications.

◆ Public parks and recreational facilities, including playground equipment, swimming pools, bath houses, tennis courts, boat docks, piers, picnic tables, and golf courses.

In order to be eligible for Public Assistance funding, the work and associated costs must:

1. Be a result of the declared event and not a pre-disaster condition.

2. Be located within the area designated by FEMA as eligible for Public Assistance.

3. Be the legal responsibility of an eligible applicant at the time of the disaster.

4. Not be under the specific authority of another Federal agency.

Additionally, eligible work is subject to applicable codes and standards and other federally mandated laws. Insurance proceeds and salvage will be deducted from the grant total when appropriate. Grants are provided to the affected State on a cost-share basis. The non-Federal contribution is shared by the State and applicant.

Other forms of Federal assistance may be available to State and local governments recovering from disaster. These programs may be activated by a presidential Major Disaster approved under the Robert T. Stafford Disaster Relief and Emergency Assistance Act, (42 U.S.C. 5121, et seq.), FEMA's authorizing legislation, or by a declaration approved under the law of another Federal agency.

Water and Disposal Systems for Rural Communities

The Farm Service Agency provides loans for installation, repair, improvement, or expansion of rural water or waste disposal systems. In some cases, grants can also be made to reduce user rates to a reasonable level for farmers, ranchers, and rural residents. This program attempts to provide basic human amenities and alleviate health hazards in rural areas, including towns of up to 10,000 inhabitants.

Health and Sanitation

The U.S. Department of Health and Human Services (HHS) may provide assistance to State and local social service agencies and to State vocational rehabilitation agencies to help them carry out emergency health and sanitation measures following a disaster. The Food and Drug Administration may work with State and local governments to establish public health controls through the decontamination or condemnation of contaminated food and drugs.

Emergency Work by the Department of Defense

During the immediate aftermath of an incident, section 403C of the Stafford Act authorizes the President to utilize personnel and equipment of the Department of Defense in certain circumstances. For instance, its resources could provide assistance in the removal of debris or in the temporary restoration of essential public facilities and services in the aftermath of a major disaster, in anticipation of a disaster declaration.

The Department of Defense can provide personnel and equipment to help restore essential public facilities and services.

The Governor of a State must request this special assistance from the Department of Defense through the FEMA Regional Director and should support the request with a finding that such work is essential for the preservation of life and property. When authorized, the work may be carried out for not more than 10 days, with the expectation that the President will issue a major disaster declaration or emergency declaration within that timeframe.

Assistance in Responding to Disasters

The U.S. Army Corps of Engineers offers special expertise in flood fighting and rescue operations. Even after floodwaters have receded, the Corps can continue to provide assistance such as debris clearance and help to restore essential public services or facilities, provided that local resources are being used to the maximum and are inadequate for the task. It can also help repair damaged flood control works or coastal protection structures.

In the event of a forest or grassland fire (either on public or private land) that becomes a major disaster, the President is authorized by the Stafford Act to provide assistance in the form of fire suppression assistance, grants, supplies, equipment, and personnel to help suppress the fire.

The U.S. Forest Service may provide personnel and equipment for search and rescue work in cooperation with State forestry agencies when the Governor requests this assistance. The agency is, of course, particularly responsible for disasters that could affect the nation's forests. The agency would provide fire protection on national forest lands and assist in controlling fires that could spread from nearby lands into national forests. The service cooperates with State foresters by providing financial and technical assistance in rural and wildland fire protection.

The U.S. Coast Guard can provide search and rescue assistance.

The U.S. Coast Guard or United States Armed Forces units may assist in search and rescue operations, in evacuating disaster victims, and in transporting supplies and equipment.

Hazard Mitigation Assistance

The Hazard Mitigation Grant Program

The Hazard Mitigation Grant Program (HMGP) can provide grants to State and local governments after a disaster has been declared. These grants provide funds to assist with the cost of mitigation measures like strengthening buildings to withstand earthquakes or raising furnaces, storage areas, or entire buildings above flood elevations. Hazard mitigation refers to measures that protect lives and property from future damages caused by natural disasters. In the long term, mitigation measures reduce personal loss, save lives, and reduce the future difficulty and cost of responding to and recovering from disasters.

Examples of types of mitigation measures eligible for HMGP funding include:

◆ Acquisition of real property in high hazard areas, demolition or relocation of structures, and conversion of land to open space use.

◆ Strengthening existing structures against high winds.

◆ Seismic rehabilitation and structural improvements to existing structures.

◆ Elevation of flood-prone structures.

◆ Implementing vegetation management programs to reduce wildfire hazard to high-risk structures.

Individuals can work with their communities to identify potential mitigation measures. The communities in a declared State can apply for HMGP funding for these measures from the State. The State is responsible for selecting and prioritizing local projects and then forwarding selected applications to FEMA for approval. The amount of funding available for the HMGP under a disaster declaration is 15 percent of FEMA's estimated total grants for all other categories of assistance from that disaster. The State sets funding priorities and allocates funds among communities. The HMGP can provide grants to assist with 75 percent of the total cost of mitigation projects. Once a project is approved, the State and local community are responsible for implementing it and providing a 25 percent funding match. This match is from State and local sources.

All mitigation projects must meet minimum eligibility criteria and comply with the National Environmental Policy Act (NEPA) and other applicable laws. HMGP funds cannot be given for acquisition or construction purposes if the project site is in an identified floodplain and the community is not participating in the National Flood Insurance Program (NFIP).

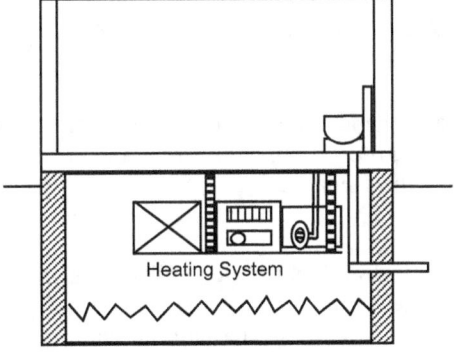

One example of an appropriate mitigation measure used in flood areas is to suspend the heating system at least 12 feet above base flood elevation.

ASSISTANCE WITHOUT A PRESIDENTIAL DISASTER DECLARATION

The preceding sections described the major types of assistance that could be made available to eligible persons, families, private businesses, and public entities after a Presidentially declared disaster. Some of this Federal assistance would be available even without a declaration. For example, assistance from the Small Business Administration, the U.S. Army Corps of Engineers, and the Department of Agriculture can be provided under the authority of their own enabling legislation. Those seeking this assistance would apply directly to these agencies.

Technical Assistance

The Federal government provides considerable technical assistance to help communities prepare for disasters and reduce loss of life and property. This assistance includes information that can help citizens assess their community's preparedness for, and vulnerability to, disasters of various types.

Programs to Prevent Floods and Protect Against Flood-Related Losses

Several Federal and State agencies play a part in providing disaster assistance for flood prevention and protection, both before and after the occurrence of a flood such as the one described in the scenario.

FEMA's Mitigation Directorate publishes maps and reports identifying flood-prone areas, flood elevations, floodways, and coastal high-hazard areas. The U.S. Geological Survey (USGS) also develops maps identifying flood-prone areas in virtually all developed and developing sections of the country. These maps can be easily obtained by contacting your district office of the U.S. Geological Survey Water Resources Division.

The U.S. Geological Survey and FEMA's Mitigation Directorate publish maps that identify flood-prone areas.

The U.S. Army Corps of Engineers can help identify areas subject to flooding by streams, lakes, and oceans. It can also provide guidance and technical services to help communities develop sound plans for land and water use that integrate knowledge of local flood hazards. By submitting a letter to the appropriate district engineer documenting the need for assistance, anyone may obtain available information. The Corps can also help communities design and construct specialized flood control projects to reduce flood damage. Any State or local agency is eligible if it has the full authority and ability to undertake the legal and financial responsibilities required for Federal participation. Applicants must submit a formal letter to the appropriate district engineer indicating clear intent to fulfill these responsibilities.

Those who live in flood-prone areas will want to know more about FEMA's National Flood Insurance Program (NFIP), administered by the Federal Insurance Administration (FIA). Most homeowners' insurance policies do not cover flooding. The NFIP enables individuals, as well as State and local governments, to purchase insurance against losses from physical damage caused by floods, flood-related mudslides, or flood-related erosion. Flood insurance claims are paid even if a disaster is not declared by the President. National Flood Insurance is available to protect buildings and/or contents in communities that have agreed to adopt and enforce sound floodplain management practices. Homeowners, business owners, and renters in a community that participates in the NFIP are eligible for flood insurance (except those areas protected by the Coastal Barrier Resource Act).

If your community is an NFIP participant, you can apply for insurance through any licensed property or casualty insurance agent or through one of the private insurance companies that are now writing flood insurance under an arrangement with the FIA. Remember, flood insurance coverage is required as a condition of receiving any Federally insured mortgage or disaster assistance for properties located within identified high risk flood-prone areas. The first step in joining the NFIP is to have a community representative contact the FEMA regional office (see Appendix A). If you want more information about flood insurance and the NFIP, you may call their toll-free number (1-800-427-4661).

Managing Land and Water Resources

The Natural Resource Conservation Service (NRCS) provides technical and financial assistance to retard runoff and prevent soil erosion. Its goal is to reduce hazards to life and property from flood, drought, and the products of erosion on any watershed impaired by a natural disaster. NRCS provides technical assistance to help rehabilitate land and

conservation systems through the Agricultural Stabilization and Conservation Service. This assistance can include cost-sharing, emergency protection against high water, and rehabilitation of rural lands damaged by natural disaster. NRCS provides information and materials (maps and reports) on watershed projects, river basin studies, and resource conservation and development areas. It also makes agency equipment available for emergency use.

The Natural Resource Conservation Service (NRCS) provides technical and financial assistance to retard runoff and prevent soil erosion.

The U.S. Forest Service also provides assistance to retard runoff and prevent soil erosion. Its goal is to safeguard life and property on, and downstream from, watershed lands suddenly damaged by fire, flood, and other natural disasters.

SUMMARY

Assistance to supplement the efforts and available resources of State and local governments is authorized under the Stafford Act when the President declares an area to be a major disaster. FEMA is authorized by the President to carry out emergency management activities at the Federal level. In addition to assistance that may be available through FEMA, a number of other types of assistance may be offered by other agencies. Some forms of Federal assistance are not dependent on a Presidential declaration. The organization that coordinates the efforts of FEMA and other agencies during a disaster is the Emergency Response Team (ERT), which is located in the affected area.

A very specific process is followed in requesting the President to declare a major disaster. It involves the cooperation and ongoing coordination of local, State, and Federal officials. On the basis of damage assessment reports and the capabilities of the local community and State to effectively respond to the disaster, the Governor makes a decision whether or not to request Federal aid. The Governor's request is submitted to the President through the FEMA Regional Director.

There are three main categories of Federal assistance—assistance for individuals and businesses, public assistance, and hazard mitigation assistance. A Presidential declaration does not guarantee that all forms of assistance will be made available; that depends on the extent and type of damage caused by the disaster, as well as the specific requests made by the Governor. ◆

Check Your Memory
(Answers on page K-2)

1. Federal assistance _____.

   a. Replaces State assistance.
   b. Supports efforts at the State and local level.
   c. Is a substitute for private insurance.
   d. Is available only under a Presidential disaster declaration.

2. In evaluating a Governor's request for Presidential disaster declaration, FEMA considers _____.

   a. Available resources of State and local governments.
   b. Imminent threats to life and safety.
   c. The State's disaster history.
   d. All of the above.

3. Other Needs Assistance under the Individuals and Households Program covers_____.

   a. Household items, furnishings, and appliances.
   b. Clothing.
   c. Privately owned vehicles.
   d. All of the above.

4. To be eligible for funding under FEMA's Public Assistance Program, disaster recovery work must _____.

   a. Be completed within 90 days of the date of declaration.
   b. Be performed by contracted labor.
   c. Be required as a direct result of a declared event.

5. All hazard mitigation projects must comply with the National Environmental Policy Act.

   a. True.
   b. False.

Unit Four

Federal Disaster

Assistance in Action

The previous unit described the kinds of help that might be available from the Federal government in the event of a disaster. This unit will provide more detailed information on how this help reaches residents of the communities that the President declares to be major disaster areas. It will explain what disaster victims should expect when applying for assistance and how they can best prepare to participate in the process.

In this unit, you will learn about:

◆ How emergency information is handled in disasters.

◆ The disaster assistance application process.

◆ The responsibilities of persons applying for assistance.

◆ How assistance is provided for communities.

EMERGENCY INFORMATION

Throughout a disaster period there is an urgent need for accurate information to reach those affected. Centerville's citizens, for example, will need to know how they can evacuate safely, where to stay, and later, where they can go for help in repairing flood-damaged homes and businesses.

Providing a uniform, coordinated, and consistent message to the public is critical before, during, and following a disaster,

As soon as the Governor is notified that the President has declared a major disaster, FEMA initiates a coordinated Federal, State, and local public affairs effort. The White House issues the initial news release announcing the declaration. FEMA issues a second release providing greater detail about counties designated and available programs. Copies are provided to the State and distributed to the media. These are the first steps in the release of information regarding the coordinated Federal and State response to the disaster.

To ensure that the public fully understands the nature of the Federal response to disasters, FEMA places a high priority on emergency information and public affairs. FEMA goes *beyond simply reacting*. The agency stresses a *proactive approach* designed to ensure the public is getting timely, accurate, consistent and easy-to-understand information from a reliable source. This approach uses all the current technological tools and requires the best available talent to reach the media and the disaster victims directly.

FEMA's approach to emergency information recognizes the importance of the partnerships with other Federal and State agencies, with local governments, and with voluntary agencies. FEMA has a unique role and an enormous responsibility when it comes to disaster assistance. It is the coordinating agency for all of the agencies that provide services during the disaster, as well as the coordinating agency for the dissemination of information.

It has been said that in times of disaster, information can be as important as food or water. Providing a uniform, coordinated, and consistent message to the public is critical. FEMA's Office of Public Affairs developed new emergency information dissemination methods to address this challenge.

The Recovery Times

The *Recovery Times* combines the latest desktop publishing technology with electronic transmission of stories and images to one printing contractor for all

disasters. Prepackaging information has facilitated quick publication and distribution of emergency information to communities. FEMA publishes the *Recovery Times* newsletter, in cooperation with State, local, and other Federal agencies, to provide timely and accurate information to disaster victims about disaster response, recovery, mitigation, and preparedness activities. Each issue contains customized content that is quickly developed for each disaster.

FEMA publishes The Recovery Times to provide timely and accurate information to disaster victims.

The publication's primary goal is to promote an understanding of disaster programs and policies— how people can apply for disaster assistance and what they can expect after they apply. Through this publication, FEMA, the State, and other government and voluntary agencies work in partnership to help disaster victims get their lives back to normal.

Recovery Times is distributed by FEMA Community Relations Representatives and is available at all disaster field offices and disaster recovery centers. It is often inserted in local newspapers and is also made available to the State and to Congressional Representatives' offices affected by the disaster.

The FEMA Radio Network

The FEMA Radio Network (FRN) is a digitized audio production and distribution system. Radio stations can call toll-free into the network 24 hours a day and obtain pre-recorded sound bites and public service announcements. FRN's state-of-the art studio supports news conferences and interviews.

FRN is easy to use. By simply dialing a toll-free telephone number, stations reach a series of recorded actualities that provide the latest up-to-the-minute reports on the Federal government's emergency response and recovery operations. Stations may then choose to record these briefs for use in their newscasts.

When a disaster hits, not only will radio listeners be completely informed on Federal emergency response activities with daily FEMA updates, but also they will hear it first-hand from FEMA officials in interviews with local newscasters.

After a disaster strikes, the recovery process starts and FRN continues to keep victims informed with information on where to apply for disaster assistance, where shelters are located, and how the disaster assistance application process works. Information is provided in other languages (such as Spanish) when the disaster area has large non-English speaking populations.

Throughout the year, FRN regularly updates its actuality service, letting radio stations know what is happening with FEMA's programs and projects around the country. FRN also provides customized public service announcements that focus on disaster threats such as hurricanes, floods, tornadoes, winter storms, earthquakes, or fires.

In addition to FRN, when situations require, a Recovery Radio Network system can be established in the disaster area. This is particularly critical in areas where communications systems have been destroyed.

The FEMA Internet Worldwide Web Site

The FEMA Internet World Wide Web site (http://www.fema.gov) is a highly popular electronic encyclopedia of disaster information. During major disasters, FEMA immediately posts a special section on that disaster and keeps it updated, by posting news releases, fact sheets, audio sound bites, and other relevant information. Real-time situation reports, maps, graphics, and links to other Internet sites with information are posted. The Web site also contains a Global Emergency Management System with links to hundreds of other emergency information sites; FEMA publications that are on-line and downloadable are listed as Resources beginning on page R-1. FEMA is committed to providing as much information as possible via this important communications medium.

In addition to FEMA's efforts, government officials at all levels will appoint public information officers (PIOs) to keep the public apprised of developments in the response and recovery effort. One of the PIO's most critical tasks at the time of a disaster is to make sure people know how to apply for assistance. The application process is started by calling a toll-free number. Disaster Recovery Centers are sometimes established to allow face-to-face interaction with program representatives.

APPLYING FOR INDIVIDUAL DISASTER ASSISTANCE

The Federal government wants to make it possible for people to get the help they need for disaster-related losses; however, disaster applicants can only receive help once for any particular loss. The Stafford Act contains a clear prohibition of any duplication of benefits. This means that if a person is compensated for disaster-related losses from any source, including private insurance, s/he cannot receive financial assistance from a Federal program for the same loss. If they receive more than one payment for the same damage, they will be required to repay the money to the Federal agency that provided the assistance. Systems are in place to detect duplication, and FEMA's Inspector General aggressively prosecutes cases of fraud and system abuse.

The majority of applications are taken by telephone through FEMA National Processing Service Centers (NPSCs). The NPSCs can take calls from anywhere in the United States and its territories. The teleregistration process takes about fifteen minutes. Individuals can help expedite the process by being prepared to provide the information that will be requested. Even though all the information may not be available, individuals should gather what they have and make the call to get the disaster application process started.

Application information to provide:

◆ Name, address of damaged property, current mailing address, and Social Security number.

◆ Telephone numbers where you can be reached.

◆ Names and ages of all persons living in the home at the time of the disaster as their primary residence.

◆ Applicant's income and the income of all other members of the household over 18 years of age.

◆ Summary of the damage.

◆ Insurance information.

As a follow-up, applicants are sent a letter from FEMA that provides a copy of the registration information, along with the names of the disaster assistance programs to which they have been referred.

What Happens Next

If a referral is made to the Individuals and Households Program (IHP), a FEMA inspector will be assigned to visit the property to verify damage and obtain information needed to determine eligibility for these programs. The inspector will contact the applicant to set up an appointment.

Once the inspector arrives at the damaged property, she or he should provide the applicant with a letter of introduction and a photo identification badge. If the inspector forgets to do this, ask to see the identification. It is always a good idea to make sure that anyone providing inspection services at your home is a legitimate inspector.

The inspector will ask the applicant to sign two documents. The first document is a certification that the information given to FEMA is true and correct; it grants to FEMA the right to use the information to determine eligibility. The second document is a declaration that the person is a United States citizen, a non-citizen national, or a qualified alien (i.e., a green-card holder). Other individuals are not eligible for Federal assistance, even if they are lawfully present in the U.S.

The inspector will ask for documentation to verify occupancy. If the applicant is a homeowner, the inspector will need documentation to verify ownership. The proof of occupancy can be a utility bill, voter registration, or statement from the landlord. The proof of ownership can be a mortgage payment book, insurance policy, or tax bill.

The inspector uses a hand-held computer to record both real and personal property damage. All aspects of the inspection are customer service-orientated, including providing sufficient time for the applicant to ask questions about disaster assistance and the inspector to answer or to provide a source for the answer.

A trained inspector makes an on-site assessment of damage on behalf of the State-administered Individual and Family Grant Program and FEMA's Disaster Housing Program.

It is important to understand the difference between the FEMA inspection and one that is done for an insurance settlement or for a Small Business Administration (SBA) loan. The FEMA inspector looks for basic needs that FEMA and the State can help with to ensure the applicant's home is a safe, sanitary, and functional place to live.

Insurance adjusters and the SBA inspectors look at all damages for purposes of providing funds to help restore the property to pre-disaster condition.

Once the FEMA inspection is complete, the information is transferred to FEMA's NPSC for eligibility determination. The applicant is notified by letter of FEMA's decision concerning assistance. If a Housing Assistance or Other Needs Assistance grant is awarded, a check is mailed to the applicant from the United States Treasury. Sometimes a State may administer the Other Needs Assistance portion of the IHP, and in this case the State would mail that part of the IHP award.

Applications to the IRS, the Red Cross, the Farm Service Agency, or other agencies will be followed up by each agency according to their own procedures.

Disaster Recovery Centers (DRC)

For some disasters, DRCs may be opened in the affected communities. The purpose of the centers is to provide a facility in the community where individuals can meet face-to-face with represented Federal, State, and local organizations and voluntary agencies to discuss their disaster-related needs and obtain information about disaster assistance programs.

Information about the locations of the DRCs and the hours of operation will be announced through the local media.

APPLYING FOR PUBLIC ASSISTANCE

The Centerville flood resulted in severe damage to many roads, bridges, buildings, utilities, and a variety of public facilities that support the community and the surrounding area. Schools, nursing homes, hospitals, and other medical care centers also incurred substantial damages. Because these facilities serve important public services, they may qualify for FEMA's Public Assistance Program.

The State is FEMA's partner in disaster recovery and works closely with the Federal government to determine how to best address community needs. Following the President's major disaster declaration, the State office of emergency management will conduct Applicants' Briefings for potentially eligible public assistance applicants. A State representative will notify the applicants of the date, time, and location of the briefing. The size of the disaster area and number of applicants will determine whether more than one briefing is held.

Inspection teams verify damages to public facilities.

The Applicants' Briefing addresses:

◆ Application procedures.

◆ Administrative requirements.

◆ Funding.

◆ Program eligibility criteria.

Applicants attending the briefing are requested to complete and submit a Request for Public Assistance (Request). The Request is an applicant's official notification of the intent to apply for public assistance. The Request outlines general information identifying the applicant, including the applicant's name, address, and primary and secondary contacts. Requests are often submitted at the Applicants' Briefing but must be submitted to the State Public Assistance Officer *within 30 days* of the declaration or designation of an area for public assistance.

An applicant should not wait until all damage is identified before requesting assistance. Federal and State personnel will review each Request to determine applicant eligibility. After an eligible applicant's Request is reviewed, the State will designate an Applicant

Liaison (Liaison) to help with assessing damages, estimating costs, and ensure that the applicant's needs are met. Similarly, once the State forwards the applicant's Request to FEMA, the applicant is assigned a Public Assistance Coordinator (PAC). The PAC is a program expert who serves as the applicant's customer service representative and works with the applicant to resolve disaster-related needs and ensures that the applicant's projects are processed as efficiently and expeditiously as possible.

Shortly after the Applicants' Briefing, the PAC will contact the applicant to schedule a Kickoff Meeting. The Kickoff Meeting is designed to provide a much more detailed review of the Public Assistance Program. The meeting is the first step in establishing a partnership between applicant, Liaison, and PAC, and is designed to focus on the specific needs of that applicant. At the meeting the PAC also discusses special considerations, such as floodplain management, insurance, hazard mitigation opportunities, and compliance with environmental and historic preservation laws, that could potentially affect the type and the amount of assistance available and the documentation needed.

A Project Worksheet (PW) is the primary tool for documenting an applicant's projects and is used to record the location of the damaged facility, damage description and dimensions, scope of eligible work, and actual or estimated costs for a project. The applicant may develop PWs for small projects (projects with an estimated cost is under $53,000 for fiscal year 2003, adjusted annually) and submit the PWs to the PAC. If the applicant submits all small project PWs to the PAC within 60 days of the Kickoff Meeting FEMA will validate 20 percent for accuracy. After 60 days FEMA will validate all PWs developed by the applicant. If the applicant requires assistance with the preparation of PWs, the PAC will assign a Project Officer or specialist to provide the applicant with technical assistance.

For large projects (projects with an estimated cost of $53,000 or greater for fiscal year 2003, adjusted annually) a Project Officer will work with the applicant to develop the PW. The Project Officer may lead a team that includes a representative of the State and one or more specialists, depending on the complexity of the project.

PWs are subject to FEMA review of cost and eligibility and to ensure compliance with FEMA's insurance and special considerations requirements. If the applicant should have any questions on the processing of his/her PWs, the applicant can contact the PAC at the Disaster Field Office, since all applications for public assistance are processed there. For more information on applying for public assistance, please refer to FEMA Publication 323, Public Assistance Applicant Handbook and the FEMA Publication 322, Public Assistance Guide.

SUMMARY

Following a Presidential declaration of a major disaster, FEMA coordinates the efforts of other Federal agencies, State and local governments, and voluntary agencies to provide disaster assistance. Public information efforts provide people with the information they need to complete the process of applying for assistance.

For individuals, families, and small businesses, the application process may be initiated through registration intake at a NPSC. Victims can facilitate the application process by carefully documenting damages and having available relevant information about themselves, their families, and businesses.

For public entities, application occurs at briefings held by State and Federal officials or by contacting the GAR. ◆

Check Your Memory
(Answers on page K-2)

1. Where would you apply for financial help to rebuild your home if it were destroyed by an earthquake? _____.

   a. At your local emergency management office.
   b. At the State emergency management office.
   c. At the local Red Cross office.
   d. Through a teleregistration process set up by FEMA.

2. When an individual requests funds through the Individuals and Households Program, a trained inspector makes an on-site inspection of the damaged property.

   a. True.
   b. False.

3. When you teleregister, you will need to provide _____.

   a. Information regarding your damages.
   b. Your social security number.
   c. The location of the damaged property.
   d. All of these.

4. Federal disaster assistance is intended to _____.

   a. Replace insurance.
   b. Help with necessary expenses not covered by insurance.
   c. Serve as the primary source of aid to disaster victims.
   d. Provide cash to victims for replacement of luxury items (such as jewelry).

5. The teleregistration process provides personnel who _____.

   a. Are available to answer your telephone calls concerning disaster assistance.
   b. Can take your application for assistance.
   c. Do both a and b.
   d. Do neither a nor b.

Unit Five

The Citizen's Role in

Disaster Preparedness

Every day, millions of people wake up, go to work, drop their children off at school, and enjoy leisure time with family and friends, following daily routines and schedules. However, when the unexpected does happen, routines change drastically, and people are suddenly aware of how fragile their lives can be. Our flood scenario demonstrated the sudden and devastating effect a disaster can have on individuals, families, and the communities in which they live.

In this unit, you will learn about:

◆ Preparedness activities that can help you and your family survive a disaster and reduce financial loss.

◆ Ways to participate in, and help improve, community preparedness.

◆ Sources of information that can help you learn more about disaster preparedness.

What people do before a disaster can make a dramatic difference in their ability to cope with and recover from a disaster, as well as their ability to protect other household members and family possessions from avoidable losses.

This unit will provide information on how individuals and families can prepare for potential disasters. Households that are prepared can reduce the fear, anxiety, and losses that surround a disaster. They can be ready to evacuate their homes, survive a period of confinement to the home, make their stays in public shelters more comfortable, and take care of their basic medical needs. They can even save each others' lives.

FINDING OUT WHAT COULD HAPPEN

The first step in preparing for any disaster is to find out which hazards could strike the community. Is the community susceptible to winter storms? Tornadoes? Earthquakes? By contacting the local emergency management office or local Red Cross office, interested individuals can find out what types of disasters are considered most likely to occur in a specific community. It is important to consider the dangers that natural hazards present when choosing a new home as well. If possible, home buyers should avoid buying in areas that are prone to floods and hurricanes.

Families should be prepared to survive emergencies likely to occur in their areas.

PROTECTING AGAINST FINANCIAL LOSS

As a protection against financial loss, homeowners should purchase insurance on their home and its contents. At a minimum, coverage should provide full replacement or replacement cost coverage. Homeowners should also investigate buying a guaranteed replacement cost policy, where available; such policies pay to rebuild a home at *today's* prices. Homes should be appraised periodically so that the policy reflects the real replacement cost.

Coverage should include special hazard-specific insurance (such as flood or earthquake insurance) appropriate for the area. Unfortunately, many homeowners learn too late that flood and earthquake loss are not covered under normal homeowners' insurance policies. Flood insurance is available in communities participating in the National Flood Insurance Program (NFIP).

Those who live in flood-prone areas in a community that is not an NFIP participant may wish to contact local officials and encourage the community to adopt the program.

Renters should purchase renter's and/or flood insurance to protect against loss for damaged or destroyed property. Be aware that the landlord's insurance will not cover damage to, or loss of, tenant's possessions.

Those concerned about their level of protection should make an appointment with their insurance agent to review current insurance coverage. It is important to get coverage early since there is usually a 30-day waiting period before it takes effect.

Insurance claims are expedited by inventories of possessions supplemented by photographs or videotape.

Any insurance claim filed will be expedited if the applicant has made an inventory of household furnishings and other possessions, supplemented with photographs or videotape. This information can be used to document property destroyed or damaged in a disaster. Computer software programs are available that can make this task less daunting. The documentation should be stored in a safe deposit box or some other safe place away from the premises. Originals of all important financial and family documents should be stored in a safe place, with copies elsewhere.

Homeowners also can take measures to protect themselves, their homes, and personal property from damage in the event of a flood, earthquake, hurricane, or other hazardous event. In flood-prone areas, homeowners can move utilities and expensive appliances such as washers and dryers to the first floor or above expected flood levels. Homeowners in California have learned to avoid placing heavy pictures above beds and to secure heavy and breakable items on shelves. Homeowners on the coast can install hurricane shutters on windows or hurricane clips to secure the roof. Protective measures can range from simple do-it-yourself activities to more expensive installations that require professional help. It is important to know the potential for a disaster event occurring near your home when deciding what types of preventative measures to undertake. The FEMA Hazard Mitigation Division offers a series of "How-to" documents that describe how to protect yourself and your property from a variety of hazards. They can be found on the FEMA website.

KNOWING THE WARNING SYSTEM

To warn their citizens in time of an emergency, some communities use sirens or loud-speakers; others rely on officials going door-to-door or on messages delivered by local TV or radio stations. The local emergency management office can provide information on what warning signals are being used in the community. It is important to know what alarms sound like, what they mean, and what action should be taken when they are heard.

NOAA Weather Radio (NWR) is a nationwide network of radio stations broadcasting continuous weather information direct from a nearby National Weather Service office. NWR broadcasts National Weather Service warnings, watches, forecasts and other hazard information 24 hours a day. Working with the Federal Communication Commission's (FCC) Emergency Alert System, NWR is an "all hazards" radio network, making it a single source for comprehensive weather and emergency information. Known as the "Voice of the National Weather Service," it is provided as a public service by the National Oceanic & Atmospheric Administration (NOAA), part of the Department of Commerce. NWR includes more than 800 transmitters, covering all 50 states, adjacent coastal waters, Puerto Rico, the U.S. Virgin Islands, and the U.S. Pacific Territories. It requires a special radio receiver or scanner capable of picking up the signal. Broadcasts are found in the public service band.

PREPARING TO EVACUATE OR STAYING AT HOME

Evacuations occur commonly throughout the United States. Hundreds of times each year, transportation or industrial accidents release harmful substances, forcing thousands of people to leave their homes and go to a safer area. More frequent causes of evacuations are fires, floods, and hurricanes. Almost every year, people in cities and communities along the Gulf and Atlantic coast evacuate in the face of approaching hurricanes. The largest peacetime evacuation occurred during Hurricane Floyd in 1999, when an estimated two million people evacuated eastern coastal areas.

The amount of time available to evacuate a home or community depends on the hazard. Sometimes, there are days to prepare: for example, hurricanes can generally be detected early. However, in sudden emergencies, such as hazardous materials spills, there may be only moments to leave the area. This means families must prepare now, because when it is time to leave home, it may be too late to collect even the most basic necessities. It also helps to consider in advance where you would go when advised to evacuate—to a designated public shelter or to relatives or friends outside the disaster area. The supplies you need should be readily available, along with a checklist to ensure that you have everything.

Evacuation periods can last for hours, several days, or even longer after a major disaster. For part, or all, of this time, citizens may be responsible for their own food, clothing, and other emergency supplies.

For some emergencies—such as winter storms or a hazardous material spill—residents may need to take shelter in their homes. Regardless of whether a safe response means evacuating or seeking shelter at home, residents should be prepared to take care of their household's needs without outside help for a minimum of 3 days. Because of the severity of damage caused by Hurricane Andrew in 1992, many families were not reached by outside help for days after the storm.

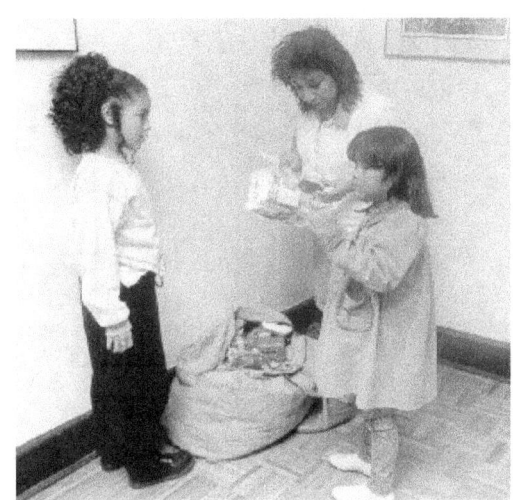

Disaster supply kits should be packed in advance to facilitate rapid evacuation.

Once a disaster is in progress, there will be no time to shop or search for supplies. But if people have gathered supplies in advance, families can endure an evacuation or home confinement. Disaster supply kits should contain the supplies listed below. The supplies should be stored in an easy-to-carry container such as a duffel bag, backpack, or covered container.

- One gallon of water per person per day, food that will not spoil, and a nonelectric can opener.
- One change of clothing and footwear, blankets, or sleeping bags.
- A first aid kit that includes the family's prescription medications.
- A battery-powered radio, a flashlight, and extra batteries.
- An extra set of car keys.
- Sanitation supplies.
- Special items for infant, elderly, or disabled family members.
- Cash and credit card.
- An extra pair of glasses.
- Matches in a waterproof container.
- Signal flare.

The kit should be kept in a convenient place near an exit door at a location known to all household members. Important family documents such as insurance policies, Social Security cards, family records, and important telephone numbers should be kept in a waterproof container in or with the disaster supplies kit. A smaller version of the kit should be kept in the trunk of the car.

It is important to maintain the supplies in the kit. The stored water supply should be changed every 3 months so it stays fresh. Food supplies should be replaced every 6 months and batteries replaced yearly. Physicians or pharmacists can provide information about storage times for prescription medications.

REUNITING AFTER A DISASTER

It is essential for household members to develop a plan for reuniting after a disaster. To prepare for a fire in the home, the family should identify a neighborhood rendezvous point located a safe distance from the house. If household members meet at the same spot, firefighters do not unnecessarily risk their lives trying to rescue someone who is already safe. This place must be designated in advance. All family members must be directed to evacuate to this designated location in the event of a fire and not to go back into a burning building.

Determining in advance where to reunite in the event of a disaster can help prevent unnecessary injury to family members or emergency workers.

For an emergency that occurs when family members are not at the same location, an out-of-state friend or relative should be asked to be the "family contact." Even when local telephone service is disrupted, long distance service often works. After a disaster, separated family members should call the family contact to let him or her know where they are. It is important to make sure everyone knows the contact's phone number.

It is also wise to know what disaster plans have been made by the children's school or day care center and where children will be sent if they are in school when an evacuation is announced. Family members should also be aware of disaster plans for places where family members work. Knowing these plans can help them find each other more easily. In case parents should become separated from their children during a disaster, they may wish to consult the doctor in advance and file a Medical Release Form to ensure that any injuries sustained by the children in a disaster would be treated promptly.

CONDUCTING PREPAREDNESS ACTIVITIES

A number of basic preparedness activities can make a dramatic difference in a family's readiness to survive and cope with a disaster.

1. Responsible household members should know where, when, and how to shut off electricity, gas, and water at main switches and valves and have the tools required to do this (usually a pipe and crescent or adjustable wrench). Taking this step can prevent dangerous leaks, explosions, and other unnecessary damage to the home. Local utility companies can provide necessary instructions. Once gas is turned off, a service representative will be required to turn it back on safely.

2. Each family member should know how to use an ABC-type fire extinguisher. The local fire department can demonstrate the proper use of extinguishers. All household residents should be shown where the extinguisher is kept. It should be tested regularly according to the manufacturer's instructions.

Responsible family members should know how to use ABC-type fire extinguishers.

3. Smoke alarms should be installed on each level of the home, especially near bedrooms. Each household should test the detector once a month and change batteries at least once or twice a year. A good time to do this is in the spring and fall when clocks are reset.

4. Even in some cases in which smoke alarms sounded, people have sometimes headed in the wrong direction in the smoke or mistakenly taken people elsewhere in the home rather than outside. It is important to plan and practice alternate escape routes. For example, is there a balcony or window in each room that could be equipped with a nearby ladder? There will not be enough time for you to give children directions if a fire occurs; it may not be possible to reach them. Therefore,

children need to know what to do on their own. Baby-sitters should also be given instructions as to alternate escape routes they and the children should use. It is important to ensure that small children can reach alternate exits. Achieving this may require a sturdy piece of furniture to be placed by the exit (usually a window) so that the child can stand on it to reach the window. Periodic fire and emergency evacuation drills are needed to practice the use of alternate exits as well as of the neighborhood rendezvous point.

5. A "home hazard hunt" should be conducted to identify objects that could block escape or cause injury if they become dislodged in an emergency. Those who live in earthquake-prone areas should remember to secure heavy objects; for example, heavy bookcases should be fastened to the wall and heavy objects must not be hung over the bed.

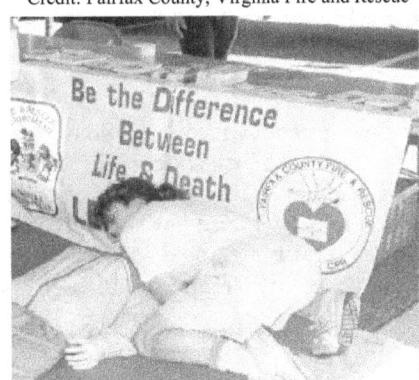

Credit: Fairfax County, Virginia Fire and Rescue

Knowing CPR or first aid can help save lives in an emergency.

6. Everyone should be encouraged to take a course on first aid and cardiopulmonary resuscitation (CPR) from the American Red Cross or other qualified sources. Knowing how to provide first aid and CPR can help save lives when immediate action is critical. Think about how frequently you are at some distance from medical help, or how difficult it would be to get treatment promptly in the first few hours or days after a major disaster.

7. Families should also consider FEMA's Community Emergency Response Team (CERT) program. It helps train people to be better prepared to respond to emergency situations in their communities. When emergencies happen, CERT members can give critical support to first responders, provide immediate assistance to victims, and organize spontaneous volunteers at a disaster site. CERT members can also help with non-emergency projects that help improve the safety of the community. The CERT course is taught in the community by a trained team of first responders who have completed a CERT Train-the-Trainer course conducted by their state training office for emergency management, or FEMA's Emergency Management Institute (EMI), located in Emmitsburg, Maryland. CERT training included disaster preparedness, disaster fire suppression, basic disaster medical operations, and light search and rescue operations.

8. Each member of the household—even children—should know how to summon help if an emergency occurs. Emergency telephone numbers should be posted by the phone (fire, emergency medical services, police, ambulance, poison control, etc.). In some areas, there is a 9-1-1 system. Everyone in the family should be prepared to provide essential information on the location and nature of the emergency.

In a disaster, volunteers come together to help their neighbors.

9. Just as a home may have hazardous areas, it will also have safe areas in which to seek shelter. The local emergency management office or the local American Red Cross chapter can provide information on safe places to seek shelter in the home. These sources will also have other materials that can help people become better prepared if disaster should strike. The Resources section of this course includes a list of emergency preparedness publications that you may obtain from FEMA and your local American Red Cross chapter.

HELPING YOUR COMMUNITY

Some of the most helpful ways people can get involved and help a community to prepare for a disaster and to respond and recover from a disaster are for them to affiliate themselves with an experienced voluntary agency through volunteering or supporting the voluntary agency with cash donations. There are many volunteer opportunities to assist in a variety of ways. See Appendix B for a summary of services provided by voluntary agencies and how to contact them.

During his 2002 State of the Union address, President George W. Bush called upon every American to get involved in strengthening our communities and sharing our compassion around the world. He called on each American to dedicate at least two years over the course of their lives to the service of others, and created the USA Freedom Corps to help answer the call. It is a coordinating council housed at the White House that is designed to help individuals find service opportunities that match their interests and talents in their hometowns, across the country or around the world. Information about Freedom Corps; its service programs under Peace Corps; Citizen Corps; Americorps; Senior Corps; and volunteer opportunities is located at http://www.usafreedomcorps.gov/.

Supporting voluntary agencies through either monetary donations or through volunteer support is in many ways the most effective way for people to get involved. The voluntary agency can provide training, guidance, and can help the volunteer find meaningful work whether it is in the disaster mitigation period or disaster preparedness, response or recovery. Cash donations to voluntary agencies help those agencies provide cash vouchers to people in need who can purchase more precisely what they need. Cash donations spent in the disaster area help contribute to bringing the local economy back to life. Cash donations also avoid the highly labor intensive tasks that most material donations require.

The collection of donated goods to support a disaster relief operation should be done only if it is in coordination with an organization that has identified a need for the goods and the donor and recipient organization are prepared to handle the shipping, receiving and distribution of the goods. Many donated goods end up being wasted because they are not the appropriate goods in the first place and little attention was paid to the logistics requirements. Even worse, donated goods coming into a disaster area often disrupt and interfere the flow of critical emergency response services.

SUMMARY

Each individual should be prepared to take appropriate protective action if a disaster should occur. This means knowing what type of disasters have the greatest chance of occurring in the local area. Understanding how the community would be warned, how to prepare the home, what supplies to stock, and how to check on family members following an emergency are all important aspects of preparedness.

There are many sources of information about emergencies and family preparedness, including the local emergency management office, FEMA, and the local American Red Cross chapter. It is important to collect information on what disasters are most likely to occur in the area and what actions should be taken in advance of or during those disasters to protect oneself.

Being prepared will decrease the chance of injury to family members and the financial loss that often results from disasters. Disaster relief can supplement individual preparedness measures—but it can never make up for a lack of planning to protect oneself and one's family. ◆

Check Your Memory
(Answers on page K-2)

1. Which of the following is the best source of information on what disasters are most likely in your area?

   a. The local police department.
   b. Your local American Red Cross chapter and emergency management office.
   c. Your Congressional representative.
   d. Your representative in your state's legislative assembly.

2. If a fire should occur in your home, all residents should plan to meet each other at what location?

   a. A specific location in the neighborhood a safe distance from your home.
   b. The fire department.
   c. The home of a friend or relative in another community.
   d. The basement of the house.

3. Which of the following requires periodic maintenance?

   a. Your smoke detector.
   b. Your fire extinguisher.
   c. The prescription medication in your disaster supplies kit.
   d. All of the above.

4. Which of the following is a true statement about sheltering in your home?

   a. You should prepare for in-house sheltering as well as for the possibility of evacuation.
   b. It is never used. You would always evacuate to ensure your safety.
   c. No particular preparation is required for in-house sheltering.
   d. In-house sheltering is always preferable to and less risky than evaluation.

5. Smoke detector batteries should be checked how often?

   a. Once a month.
   b. Once or twice a year.
   c. Every 2 years.
   d. Whenever you think about it.

APPENDIX A
STATES AND TERRITORIES ASSIGNED TO FEMA REGIONAL OFFICES

Region I (based in Boston, MA) Includes Connecticut, Maine, Massachusetts, New Hampshire, Rhode Island, and Vermont

Region II (based in New York, NY) Includes New Jersey, New York, Commonwealth of Puerto Rico, and U.S. Virgin Islands

Region III (based in Philadelphia, PA) Includes Delaware, the District of Columbia, Maryland, Pennsylvania, Virginia, and West Virginia

Region IV (based in Atlanta, GA) Includes Alabama, Florida, Georgia, Kentucky, Mississippi, North Carolina, South Carolina, and Tennessee

Region V (based in Chicago, IL) Includes Illinois, Indiana, Michigan, Minnesota, Ohio, and Wisconsin

Region VI (based in Denton, TX) Includes Arkansas, Louisiana, New Mexico, Oklahoma, and Texas

Region VII (based in Kansas City, MO) Includes Iowa, Kansas, Missouri, and Nebraska

Region VIII (based in Denver, CO) Includes Colorado, Montana, North Dakota, South Dakota, Utah, and Wyoming

Region IX (based in Oakland, CA) Includes Arizona, California, Hawaii, Nevada, American Samoa, Guam, Commonwealth of the Northern Mariana Islands, Commonwealth of Federated States of Micronesia, Republic of the Marshall Islands, and Palau

Region X (based in Bothell, WA) Includes Alaska, Idaho, Oregon, and Washington

APPENDIX B

VOLUNTARY AGENCIES
ACTIVE IN THE UNITED STATES

The following agencies often play such a major role in disasters that a brief description is in order.

The *Adventist Community Services* (ACS) receives, processes, and distributes clothing, bedding, and food products. in major disasters, the agency brings in mobile distribution units filled with bedding and packaged clothing that is presorted according to size, age, and gender. ACS also provides emergency food and counseling and participates in the cooperative disaster child care program.

The *American Radio Relay League, Inc.* (ARRL) is a national volunteer organization of licensed radio amateurs in the United States. ARRL-sponsored Amateur Radio Emergency Services (ARES) provide volunteer radio communications services to Federal, State, county, and local governments, as well as to voluntary agencies. When telephone and power grids are knocked out, members volunteer not only their services but also their privately owned radio communications equipment.

The *American Red Cross* is required by Congressional charter to undertake disaster relief activities to ease the suffering caused by a disaster. Emergency assistance includes fixed/mobile feeding stations, shelter, cleaning supplies, comfort kits, first aid, blood and blood products, food, clothing, emergency transportation, rent, home repairs, household items, crisis intervention, and medical supplies. Additional assistance for long-term recovery may be provided when other relief assistance and/or personal resources are not adequate to meet disaster-caused needs. The American Red Cross provides referrals to the government and other agencies providing disaster assistance.

The *America's Second Harvest* collects, transports, warehouses, and distributes donated food and grocery items for other charitable agencies involved in direct feeding operations and the distribution of relief supplies through its network of food banks and food rescue organizations. America's Second Harvest also processes products collected in food drives by communities wishing to help the disaster-affected community. America's Second Harvest is fully domestic in scope and practice.

The *Ananda Marga Universal Relief Team* (AMURT) renders immediate medical care, food and clothing distribution, stress management, and community and social services. AMURT also provides long-term development assistance and sustainable economic programs to help disaster-affected people. AMURT depends primarily on full- and part-time volunteer help, and has a large volunteer base to draw on worldwide. AMURT provides and encourages disaster services training in conjunction with other relief agencies like the American Red Cross.

The *Catholic Charities USA* Disaster Response is the organization that unites the social services agencies operated by most of the 175 Catholic dioceses in the United States. The Disaster Response section of Catholic Charities USA provides assistance to communities in addressing the crisis and recovery needs of local families. Catholic Charities agencies emphasize ongoing and long-term recovery services for individuals and families, including temporary housing assistance for low income families, counseling programs for children and the elderly, and special counseling for disaster relief workers.

The *Christian Disaster Response* (CDR) worked in cooperation with the American Red Cross, the Salvation Army, Church World Service Disaster Response, and NOVAD to enable local church members to become effective volunteers for assignment on national disasters. CDR provides disaster assessments, fixed/mobile feeding facilities, and in-kind disaster relief supplies. CDR also coordinates and stockpiles the collection of donated goods through their regional centers throughout the U.S.

The *Christian Reformed World Relief Committee* (CRWRC) has the overall aim of assisting churches in the disaster-affected community to respond to the needs of persons within that community. CRWRC provides advocacy services to assist disaster victims in finding permanent, long-term solutions to their disaster-related problems, as well as housing repair and construction, needs assessment, clean-up, child care, and other recovery services.

The *Church of the Brethren Emergency Response/Service Ministries* supports disaster response by sending teams of skilled volunteers to clean up debris and repair or rebuild homes for disaster survivors in need. Disaster Child Care (DCC) provides specially trained volunteers to comfort and care for traumatized children at disaster sites and following aviation incidents. DCC volunteers come from many different denominations.

The *Church World Service* (CWS) Emergency Response Program builds structure to assist disaster survivors and promotes mitigation and disaster planning/training in the United States on behalf of its 36-member communions and other affiliated agencies. CWS Disaster Response and Recovery Liaisons (DRRLs) foster and support the development of community-based disaster recovery organizations. In disaster response, upon request CWS will provide direct material support such as blankets and health, school, clean-up, and infant supplies.

The *Episcopal Church Presiding Bishop's Fund for World Relief* responds to domestic disasters principally through its network of nearly 100 U.S. dioceses and over 8,200 parishes. It also sends immediate relief grants for such basics as food, water, medical assistance, and financial aid within the first 90 days following a disaster. Ongoing recovery activities are provided through rehabilitation grants, which offer the means to rebuild, replant ruined crops, and counsel those in trauma. The Episcopal Church works primarily through Church World Service in providing its disaster-related services.

The *Friends Disaster Service* (FDS) provides clean-up and rebuilding assistance to the elderly, disabled, low income, or uninsured survivors of disasters. It also provides an outlet for Christian service to Friends' volunteers, with an emphasis on love and caring. In most cases, FDS is unable to provide building materials and, therefore, looks to other NVOAD member agencies for these materials.

The *Humane Society of the United States* (HSUS) supports the Nation in times of emergency with its comprehensive approach to protecting animals in disaster through disaster response and rescue training as well as preparedness education and community planning. The HSUS provides needed support directly to communities during disasters through its Disaster Animal Response Teams (DARTs) that provide rescue, sheltering, and first aid and evaluation of animals, donation and volunteer management, and incident management. The HSUS also offers practical information for emergency managers, voluntary agencies, and animal care and control agencies about disaster plans that include animals.

The *International Association of Jewish Vocational Services* (IAJVS) is an affiliation of 26 U.S., Canadian, and Israeli Jewish Employment and Vocational and Family Services agencies that provides a broad spectrum of training and employment initiatives needed in disaster. Some of these specific services include vocational evaluation, career counseling, skills training, and job placement. In addition to providing vocational services, IAJVS is also involved in problems of drug and alcohol abuse programs for the homeless, specialized services for welfare recipients, and workshops for disabled individuals.

The *International Relief Friendship Foundation, Inc.* (IRFF) provides disaster assistance through three main strategies. First, IRFF conducts an assessment of needs/unmet needs through a systematic canvassing of affected communities and neighborhoods. Second, IRFF provides appropriate spiritual care to all victims of disaster through personnel trained in disaster care and relief. Third, IRFF provides volunteer support for the initial clean-up phase of the disaster through its network of individuals, families, and independent clergy and their congregations. IRFF places particular emphasis on those sectors of a community that may be hidden or unnoticed and therefore unaware of how to access disaster relief.

The *Lutheran Disaster Response* (LDR) provides for immediate disaster response, in both natural and technological disasters, long-term rebuilding efforts, and support for preparedness planning through synods, districts, and social ministry organizations. The disasters to which LDR responds are those in which needs outstrip available local resources. LDR provides for the coordination of 6,000 volunteers annually. In addition, LDR provides crisis counseling, support groups, mental health assistance, and pastoral care through its accredited social service agencies.

Mennonite Disaster Services assists disaster victims by providing volunteer personnel to clean up and remove debris from damaged and destroyed homes and personal property and to repair or rebuild homes. Special emphasis is placed on assisting those less able to help themselves, such as the elderly and handicapped.

The *National Emergency Response Team* (NERT) meets the basic human needs of shelter, food, and clothing during times of crisis and disaster. NERT provides Emergency Mobile Trailer units (EMTUs), which are self-contained, modest living units for up to 8-10 people, to places where disaster occurs. When EMTUs are not in use, they serve as mobile teaching units used in Emergency Preparedness programs in communities.

The *National Organization for Victim Assistance* provides social and mental health services for individuals and families who experience major trauma after disaster, including critical incident debriefings.

The *Nazarene Disaster Response* provides clean-up and rebuilding assistance, especially to the elderly, disabled, widowed, and those least able to help themselves. In addition, a National Crisis Counseling Coordinator works into the recovery phase by assisting with the emotional needs of disaster victims.

The *Northwest Medical Teams International* supports the lead voluntary agencies responding to disaster situations by enlisting volunteers as needed to the stricken areas and sending money and supplies for cleaning and reconstruction.

The *Phoenix Society for Burn Survivors* is a non-profit 501(c)(3) organization whose mission is to provide support to those affected by a burn injury through peer support, education, collaboration, and advocacy. The Phoenix Society's goal is to assure that all burn survivors have access to the support systems necessary for their recovery. The Society encourages and supports activities that bring survivors together to learn and grow from their common experiences.

The *Points of Light Foundation* coordinates spontaneous, unaffiliated volunteers and meets the needs of the local community and other disaster response agencies through its affiliated network of local Volunteer Centers.

The *Presbyterian Disaster Assistance* (PDA) is the crisis ministry of the Presbyterian Church (USA). PDA seeks to work cooperatively with other voluntary organizations through NVOAD. When a major disaster occurs, response by PDA may include the following: human resources, volunteer clean-up and rebuilding teams, financial resources, material resources, food and shelter, and pastoral care.

The *REACT International* provides emergency communication facilities for other agencies through its national network of Citizens Band radio operators and volunteer teams. REACT teams are encouraged to become part of their local disaster preparedness plan. Furthermore, they are encouraged to take first aid training and to become proficient in communications in time of disaster.

The *Salvation Army* provides emergency assistance including mass and mobile feeding, temporary shelter, counseling, missing person services, medical assistance, and distribution of donated goods including food, clothing, and household items. It also provides referrals to government and private agencies for special services.

The *Second Harvest National Network of Food Banks* collects, transports, warehouses, and distributes donated food and grocery items for other charitable agencies involved in direct feeding operations and the distribution of relief supplies through its network of food banks and food rescue organizations. America's Second Harvest also processes products collected in food drives by communities wishing to help the disaster-affected community. America's Second Harvest is fully domestic in scope and practice.

The *Society of St. Vincent De Paul* provides social services to individuals and families, and collects and distributes donated goods. It operates retail stores, homeless shelters, and feeding facilities that are similar to those run by the Salvation Army. The stores' merchandise can be made available to disaster victims. Warehousing facilities are used for storing and sorting donated merchandise during the emergency period.

The *Southern Baptist Disaster Relief* provides more than 200 mobile feeding units staffed by volunteers who can prepare and distribute thousands of meals a day. Active in providing disaster childcare, the agency has several mobile childcare units. Southern Baptists also assist with clean-up activities, temporary repairs, reconstruction, counseling, and bilingual services.

The *UJA Federations of North America* organizes direct assistance, such as financial and social services, to Jewish and general communities in the U.S. following disaster. It also provides rebuilding services to neighborhoods and enters into long-term recovery partnerships with residents.

The *United Methodist Committee on Relief* provides funding for local units in response and recovery projects based on the needs of each situation. This agency also provides spiritual and emotional care to disaster victims and long-term care of children impacted by disaster.

The *United States Service Command* provides trained corps of volunteers to voluntary and governmental agencies during disaster.

The *Volunteers in Technical Assistance* provides telecommunications and management information systems support to the emergency management community.

The *Volunteers of America* is involved in initial response services aimed at meeting the critical needs of disaster victims, such as making trucks available for transporting victims and supplies to designated shelters. It also collects and distributes donated goods and provides mental health care for survivors of disaster.

The *World Vision* procures high quality gifts-in-kind throughout the United States, and distributes those goods to disaster survivors through community distribution centers. It also identifies under-served neighborhoods and works with unaffiliated faith-based organizations to mobilize response efforts by providing training and cash grants.

NVOAD Membership

Adventist Community Services
12501 Old Columbia Pike
Silver Spring, MD 20904-1608

The American Radio Relay League, Inc.
Administrative Headquarters
225 Main Street
Newington, CT 06111

American Red Cross Disaster Services
8111 Gatehouse Road, Second Floor
Falls Church, VA 22042

America's Second Harvest
35 East Wacker Drive, Suite 2000
Chicago, IL 60601

Ananda Marga Universal Relief Team (AMURT)
6810 Tilden Lane
Rockville, MD 20852
(301) 984-0217
(301) 984-0218 (fax)

Catholic Charities, USA
1731 King Street, Suite 200
Alexandria, VA 22314

Christian Disaster Response
P.O. Box 3339
Winter Haven, FL 33885-3339

Church of the Brethren General Board
P.O. Box 188, 601 Main Street
New Windsor, MD 21776

Church World Service
Emergency Response Program
475 Riverside Drive, No. 700
New York, NY 10115

Christian Reformed World Relief Committee
2850 Kalamazoo Avenue, SE.
Grand Rapids, MI 49560

The Episcopal Church
The Presiding Bishop's Fund for World Relief
815 Second Avenue
New York, NY 10017

Friends Disaster Services (Quakers)
241 Keenan Road
Peninsula, OH 44264

Humane Society of the United States
2100 L Street, NW.
Washington, DC 20037
(202) 452-1100
disaster@hsus.org

International Association of Jewish Vocational Services (IAJVS)
1845 Walnut Street, Suite 608
Philadelphia, PA 19103

International Relief Friendship Foundation
177 White Plains Road, No. 50F
Tarrytown, NY 10591

Lutheran Disaster Response
8765 West Higgins Road
Chicago, IL 60631

Mennonite Disaster Services
21 South 12th Street
P.O. Box 500
Akron, PA 17501

National Emergency Response Team (NERT)
1058 Albion Road
Unity, ME 04988

National Organization for Victim Assistance
1757 Park Road, NW.
Washington, DC 20010

Nazarene Disaster Response
P.O. Box 585186
Orlando, FL 32858-5186

Northwest Medical Teams International
P.O. Box 10
Portland, OR 97207-0010

NOVAD Staff
7213 Central Avenue
Takoma Park, MD 20912

The Phoenix Society for Burn Survivors
Amy Acton, RN, BSN, Executive Director
2153 Wealthy Street SE., Suite 215
East Grand Rapids, MI 49506

The Points of Light Foundation
1400 Eye Street, NW.
Suite 800
Washington, DC 20005-2208

Presbyterian Disaster Assistance
100 Witherspoon Street
Louisville, KY 40202-1396

REACT International
630 Washington Street
Allentown, PA 18102-1606

The Salvation Army
615 Slaters Lane
P.O. Box 269
Alexandria, VA 22313

Southern Baptist Convention
N.A. Mission Board
4200 North Point Parkway
Alpharetta, GA 30022-4176

St. Vincent de Paul
58 Progress Parkway
St. Louis, MO 63043-3706

UJA Federation of North America
1750 Euclid Avenue
Cleveland, OH 44115

United Methodist Committee on Relief
1601 North Kent Street, Suite 902
Arlington, VA 22209

United Methodist Committee on Relief
4630 Holston Drive
Knoxville, TN 37914

United States Service Command
P.O. Box 1084
North Chicago, IL 60064

United States Service Command
33 Penwood Drive
Whiting, NJ 08759-2057

Volunteers In Technical Assistance
1600 Wilson Blvd., Suite 710
Arlington, VA 22209

Volunteers of America
110 South Union Street
Alexandria, VA 23214

World Vision
P.O. Box 9716
Federal Way, WA 98063-9716

GLOSSARY

Declaration

The President's decision to make Federal assistance available under the Stafford Act.

Disaster Field Office (DFO)

The office established in or near the designated area to support Federal and State response and recovery operations. The DFO houses the Federal Coordinating Officer, the Emergency Response Team, and where possible, the State Coordinating Officer and support staff.

Disaster Recovery Center (DRC)

A temporary facility where, under one roof, local and State governments and voluntary agencies provide information about disaster assistance programs.

Emergency Operations Plan or Emergency Response Plan

A document that contains information on the actions that may be taken by a governmental jurisdiction to protect people and property before, during, and after a disaster.

Federal Coordinating Officer (FCO)

The person appointed by the FEMA Director, or in his/her absence, the FEMA Deputy Director, or alternatively the FEMA Associate Director for Response and Recovery, following a declaration of a major disaster or of an emergency by the President, to coordinate Federal assistance.

Federal Emergency Management Agency (FEMA)

The Federal Emergency Management Agency manages the President's Disaster Relief Fund and coordinates the disaster assistance activities of all Federal agencies in the event of a Presidential disaster declaration.

Federal Response Plan

The FRP is the Federal government's plan of action for assisting affected States and local jurisdictions in the event of a major disaster or emergency. As the implementing document for the Stafford Act, the FRP organizes the Federal response by grouping potential response requirements into 12 functional categories, called Emergency Support Functions. The FRP was completed in April 1992, revised in April 1999, and 27 Federal departments and agencies are signatories to the plan.

FEMA-State Agreement

A formal legal document between FEMA and the affected State that describes the understandings, commitments, and binding conditions for assistance applicable as a result of a declaration by the President. It is signed by the FEMA Regional Director and the Governor.

Governor

As defined by the Stafford Act, the chief executive of any State.

Grant

Financial aid given by certain Federal, State, and private agencies to help meet disaster-related necessary expenses or needs when affected individuals cannot meet such expenses or needs through insurance or other means.

Helpline

Toll-free telephone services established by FEMA that provide help to an individual to determine the status of his or her application for assistance.

Individual Assistance	Supplementary Federal assistance provided under the Stafford Act to individuals and families adversely affected by a major disaster or an emergency. Such assistance may be provided directly by the Federal government or through State and local governments or disaster relief organizations.
Major Disaster	As defined in the Stafford Act, "Any natural catastrophe (including any hurricane, tornado, storm, high water, wind-driven water, tidal wave, earthquake, tsunami, volcanic eruption, landslide, mudslide, snowstorm, or drought), or, regardless of cause, any fire, flood, or explosion in any part of the United States, which in the determination of the President causes damage of sufficient severity and magnitude to warrant major disaster assistance under this Act to supplement the efforts and available resources of States, local governments, and disaster relief organizations in alleviating the damage, loss, hardship, or suffering caused thereby."
National Processing Service Center (NPSC)	The National Processing Service Centers (NPSCs) perform a range of functions related to application processing, verification, and close-out. Key among these functions is teleregistration – contact with disaster-affected individuals via telephone for the purpose of identifying needs, determining

eligibility for assistance, and completing an application for assistance. The NPSCs also schedule inspectors to verify damage, and manage assistance records associated with an incident.

Preliminary Damage Assessment (PDA)

A process used to determine the impact and magnitude of damage and the resulting unmet needs of individuals, businesses, the public sector, and the community as a whole. Information collected as a result of the PDA process is used by the State as a basis for the Governor's request for Federal assistance under the Stafford Act, and by FEMA to document the recommendation made to the President in response to the Governor's request.

Preparedness

Those activities, programs, and systems that exist prior to an emergency that are used to support and enhance response to an emergency or disaster.

Public Assistance (PA)

Supplementary Federal assistance provided under the Stafford Act to State and local governments or certain private, nonprofit organizations other than assistance for the direct benefit of individuals and families. Public assistance also includes Community Disaster Loans and Fire Suppression Grants.

Stafford Act

The Robert T. Stafford Disaster Relief and Emergency Assistance Act, which provides the greatest single source of Federal disaster assistance.

State Coordinating Officer (SCO)   The person appointed by the Governor, upon a declaration of a major disaster or of an emergency, to coordinate State and local disaster assistance efforts with those of the Federal government, and to act in cooperation with the Federal Coordinating Officer to administer disaster recovery efforts.

RESOURCES

FEDERAL EMERGENCY MANAGEMENT AGENCY

The following publications are available without charge from your local or State emergency management office or by writing to the Federal Emergency Management Agency, P.O. Box 2012, Jessup, Maryland 20794-2012. Please refer to title and number when ordering.

General Emergency Preparedness

Order #	FEMA #	Title
8-0908	H-34	Are You Ready? Your Guide to Disaster Preparedness
8-1108	K-81	The Good Ideas Book
8-0963	L-154	Emergency Preparedness Checklist
8-1026	L-154S	Emergency Preparedness Checklist (available in Spanish)
8-1017	L-154M	Emergency Checklist/Mobility Impaired
8-0941	L-189	Family Disaster Supplies Kit
8-1004	L-189S	Family Disaster Supplies Kit (available in Spanish)
8-0954	L-191	Family Disaster Plan
8-0996	L-191S	Family Disaster Plan (available in Spanish)
8-1034	L-196	Helping Children Cope with Disaster
9-0044	L-210	Food and Water Supplies in an Emergency
0-7045	FEMA-20	FEMA Publications Catalog
8-0628	FEMA-141	Emergency Management Guide for Business and Industry
8-0958	FEMA-218	Preparedness for Hazardous Materials Emergencies in Railyards: Guidance for Railroads and Adjacent Communities
8-1123	FEMA-243	Disaster Preparedness Coloring Book
9-1125	FEMA-291	Before Disaster Strikes---How to make sure you're *financially* prepared to deal with a natural disaster
9-1124	FEMA-292	After Disaster Strikes---How to make sure you're *financially* prepared to deal with a natural disaster

9-1138	FEMA 292S	After Disaster Strikes---How to recover *financially* from a natural disaster
9-1137	FEMA 291S	After Disaster Strikes---How to recover *financially* from a natural disaster
8-0991		Hazardous Materials Exercise Evaluation Methodology (HM-EEM)
8-0601	NRT-1	Hazardous Materials Emergency Planning Guide
8-0693	NRT-1A	Criteria for Review of Hazardous Materials Emergency Plans
9-1051	SLG-101	Guide for All-Hazard Emergency Operations Planning

Winter Storms

Order #	FEMA #	Title
5-0031	L-97	Winter Fire Safety Tips for the Home

Floods

Order #	FEMA #	Title
8-0373	FEMA-55	Coastal Construction Manual
9-1905	FEMA-387	Answers to Questions About the National Flood Insurance Program
8-0383	DAP-16	When You Return to a Storm-Damaged Home

Fire

Order #	FEMA #	Title
5-0172	FA-246	Escape from Fire: Once You're Out, Stay Out
5-0228	L-203	Wildfire - Are You Prepared?

Dam Safety

Order #	FEMA #	Title
8-0459	FEMA-64	Emergency Action Planning Guidelines for Dams

Hurricanes

Order #	FEMA #	Title
0-0017	L-105	Safety Tips for Hurricanes
8-0373	FEMA-55	Coastal Construction Manual
8-0900	—	Hurricane Wallet Card (English)
8-0901	—	Hurricane Wallet Card (Spanish)
8-0950	—	Video—Hurricane: It's Not Just Another Storm
8-0440	L-212	Hurricane Awareness—Action Guidelines for Senior Citizens
8-0864	L-107	Hurricane - Floods—Safety Tips for Coastal and Inland Flooding
8-1091	L-204	Hurricane Safety Tips

Tornadoes

Order #	FEMA #	Title
0-0164	L-148	Tornado Safety Tips Fact Sheet
9-1363	FEMA 320	Taking Shelter from the Storm; Building a Safe Room Inside Your Home

Earthquakes

Order #	FEMA #	Title
8-0750	FEMA-48	Coping with Children's Reactions to Earthquakes and Other Disasters (English)
8-0487	FEMA-75	Preparedness for People with Disabilities (Earthquake Preparedness)
8-0488	FEMA-76	Preparedness in High-Rise Buildings (Earthquake Preparedness)
8-0821	L-111	Earthquake Safety Tips
2-0006	L-143	Earthquake Preparedness in Apartments and Mobile Homes
2-0007	FEMA-46	Earthquake Safety Checklist

PUBLICATIONS JOINTLY PRODUCED BY FEMA AND THE AMERICAN RED CROSS

Order #	FEMA #	Red Cross #	Title
8-0954	FEMA L-191	ARC 4466	Your Family Disaster Plan
8-0996	FEMA L-191S	ARC 4466S	Su Plan Para el Hogar en Caso de Desastres
8-0963	FEMA L-154	ARC 4471	Emergency Preparedness Checklist
8-0941	FEMA L-189	ARC 4463	Your Family Disaster Supplies Kit

PUBLICATIONS JOINTLY PRODUCED BY THE HUMANE SOCIETY OF THE UNITED STATES AND THE AMERICAN RED CROSS

Order #	HSUS #	Red Cross #	Title
	HSUS PM2161	ARC 321355	Pets & Disasters: Get Prepared
	HSUS GR3244	ARC 657100	Pets First Aid

PUBLICATIONS JOINTLY PRODUCED BY THE NATIONAL ENDOWMENT FOR FINANCIAL EDUCATION AND THE AMERICAN RED CROSS

Order #	FEMA #	Red Cross #	Title
9-1124	FEMA 291	ARC 5076	After Disaster Strikes...How to Recover Financially From a Natural Disaster
9-1125	FEMA 291	ARC 5075	Before Disaster Strikes...How to Make Sure You're Financially Prepared to Deal with a Natural Disaster (also available in Spanish)

AMERICAN RED CROSS

The American Red Cross publishes many materials for use in helping the public prepare for and respond appropriately when disaster strikes. The publications cover general topics of family disaster preparedness, as well as specific hazards such as hurricanes, floods, tornadoes, and terrorism. A complete index of all available materials can be found on the Red Cross web site by visiting www.redcross.org/pubs/dspubs/cde.html. Your local American Red Cross chapter also may have developed publications specifically tailored to your area. Check with your local chapter for available publications.

ANSWERS TO PRETEST AND CHECK YOUR MEMORY

PRETEST

1.	c	(Material covered in Unit One)
2.	a	(Material covered in Unit One)
3.	b	(Material covered in Units One, Two, and Three)
4.	b	(Material covered in Units One and Four)
5.	b	(Material covered in Units One and Three)
6.	b	(Material covered in Units One and Four)
7.	d	(Material covered in Unit Two)
8.	a	(Material covered in Unit Two)
9.	a	(Material covered in Unit Two)
10.	a	(Material covered in Unit Two)
11.	b	(Material covered in Unit Three)
12.	d	(Material covered in Unit Three)
13.	c	(Material covered in Unit Four)
14.	d	(Material covered in Unit Three)
15.	a	(Material covered in Unit Three)
16.	a	(Material covered in Unit Three)
17.	c	(Material covered in Unit Three)
18.	c	(Material covered in Unit Four)
19.	b	(Material covered in Unit Three)
20.	a	(Material covered in Unit Four)
21.	c	(Material covered in Unit Three)
22.	a	(Material covered in Unit Five)
23.	b	(Material covered in Unit Five)
24.	d	(Material covered in Unit Five)
25.	a	(Material covered in Unit Five)

CHECK YOUR MEMORY

Unit One

1.	c	(See page 1-8)
2.	d	(See page 1-9)
3.	a	(See page 1-8)
4.	b	(See page 1-10)
5.	b	(See page 1-10)

Unit Two

1.	d	(See pages 2-1 and 2-2)
2.	b	(See pages 2-4 and 2-5)
3.	d	(See page 2-6)
4.	d	(See pages 2-9)
5.	b	(See page 2-9)
6.	b	(See page 2-5)

Unit Three

1.	b	(See page 3-2)
2.	d	(See pages 3-10 and 3-11)
3.	d	(See page 3-17)
4.	c	(See page 3-24)
5.	a	(See page 3-27)

Unit Four

1.	d	(See page 4-6)
2.	a	(See pages 4-6 and 4-7)
3.	d	(See page 4-6)
4.	b	(See page 4-5)
5.	c	(see page 4-6)

Unit Five

1.	b	(See page 5-2)
2.	a	(See page 5-7)
3.	d	(See pages 5-6 to 5-10)
4.	a	(See pages 5-1 to 5-3)
5.	a	(See page 5-8)

FINAL EXAMINATION

HOW TO TAKE THE FINAL EXAMINATION

The following final examination is designed to find out how much you have learned about disaster assistance from this course.

While taking the final examination, read each question carefully and select the answer that you think is correct after reading all the possible choices. Complete all of the questions. You may refer to the course materials to help you answer the questions.

When you have answered all of the questions, you may submit your answers online at http://training.fema.gov, click on FEMA Independent Study and follow the link to the specific course or you may order an Opscan Answer Sheet by going to http://training.fema.gov,click on FEMA Independent Study and go to Opscan Request. Follow the instructions on the answer sheet an mail it to:

FEMA Independent Study Program
16825 South Seton Avenue
Emmitsburg, MD. 21727

Your answers will be scored and the results will be issued to you. If you score 75 percent or higher a certificate of completion will be issued to you. If you score less than 75 percent, you will be given another chance to take the test.

The final examination consists of 50 questions and should take no more than 60 minutes to complete. Find a quiet spot where you will not be interrupted during this time.

FINAL EXAMINATION

A CITIZEN'S GUIDE TO DISASTER ASSISTANCE

There is only one correct answer for each question. When you have finished, prepare the answer sheet as directed and mail to the address provided or you may submit your anwers online at http://training.fema.gov, click on FEMA Independent Study and follow the links to the specific course. Your examination will be evaluated and the results will be issued to you.

1. The natural events that most frequently result in the loss of lives and property are _____.

    a. Floods.
    b. Volcanoes.
    c. Tornadoes.
    d. Landslides.

2. The _____ provides warnings on potentially hazardous weather conditions as they develop.

    a. Federal Emergency Management Agency.
    b. National Weather Service.
    c. Local emergency management office.
    d. Skywarn agency.

3. Planning should occur prior to an emergency in order to lessen its effects.

    a. True
    b. False

4. The period in which actions taken to repair damages, alleviate disruption from a disaster, and facilitate the return to normal is called

    _____.

    a. Hazard mitigation.
    b. Disaster planning.
    c. Disaster response.
    d. Disaster recovery.

5. If disaster assistance were represented as a pyramid, the bottom of the pyramid—the most common source of disaster assistance—would be

   _____.

   a. The federal government.
   b. The State government.
   c. Local government and private agencies.
   d. International Relief Agencies.

6. Because federal assistance would be available in the event of a serious disaster, insurance is not needed.

   a. True
   b. False

7. Any event that results in significant harm to multiple lives and properties, as well as disruption to normal patterns of living, may be called a

   _____.

   a. Hazard.
   b. Disaster.
   c. Mitigation.
   d. Declaration.

8. FEMA and the federal government would assume total responsibility for disaster recovery in a Presidentially declared disaster.

   a. True
   b. False

9. Most financial assistance from the federal government is in the form of loans.

   a. True
   b. False

10. Federal funds received to repair a roof should not be used to address other needs, such as replacing damaged carpet.

    a. True
    b. False

11. The first line of defense against emergencies—and the entity primarily responsible for emergency response—is the _____.

    a. Federal government.
    b. State government.
    c. National Weather Service.
    d. Local government.

12. Measures that help restore essential services immediately following a disaster so the community can reach minimum operating standards are part of _____.

    a. Hazard mitigation.
    b. Hazard preparedness.
    c. Short-term recovery.
    d. Long-term recovery.

13. Mutual aid agreements can _____.

    a. Facilitate assistance from neighboring communities.
    b. Prevent disasters.
    c. Increase property values.
    d. Eliminate the need for insurance.

14. Situation reports are used to _____.

    a. Convey information about an emergency and possible resource needs.
    b. Report on a community's status in the National Flood Insurance Program.
    c. Replace electronic media if power fails.
    d. Provide updates on Disaster Recovery Centers.

15. The State office that coordinates deployment of State personnel and resources is the _____.

    a. Department of Human Services.
    b. National Guard.
    c. Emergency Management Office.
    d. Department of Agriculture.

16. If a State declares an emergency, the _____ is usually the individual legally responsible for mobilizing State resources.

    a. Director of the National Guard.
    b. Governor.
    c. Director of the department of public safety.
    d. Local elected official.

17. What State agency or agencies typically assist voluntary agencies such as the American Red Cross in their efforts to provide relief to victims?

    _____.

    a. National Guard.
    b. Social service agencies.
    c. Natural resources agencies.
    d. Agriculture departments.

18. A request for a Presidential declaration for a disaster comes from the

    _____.

    a. Local elected official.
    b. State emergency manager.
    c. FEMA Director.
    d. Governor.

19. The typical State emergency response plan is similar in structure and organization to most emergency operations plans developed by local governments.

    a. True
    b. False

20. State personnel play no role in situation monitoring or any other disaster-related function unless there is a State declaration of emergency.

    a. True
    b. False

21. Until 1950, Congress had to pass a separate law to provide federal funds for each major disaster that occurred.

    a. True
    b. False

22. A hospital damaged in a Presidentially declared disaster might receive aid through what category of federal assistance? _____.

    a. Hazard Mitigation Assistance.
    b. Public Assistance.
    c. Community Assistance.
    d. Individual and Business Assistance.

23. Today, the federal government's legislative authority to provide relief in a major disaster stems from what Act? _____.

    a. The Comprehensive Disaster Assistance Act.
    b. The National Flood Insurance Act.
    c. The National Security Emergency Preparedness Act.
    d. The Robert T. Stafford Disaster Relief and Emergency Assistance Act.

24. Debris removal, search and rescue, and demolition of buildings that immediately threaten public safety are examples of _____ under FEMA's Public Assistance Program.

    a. Permanent work.
    b. Ephemeral work.
    c. State work.
    d. Emergency work.

25. A small business damaged in a Presidentially declared disaster might receive aid through what category of federal assistance? _____.

    a. Public Assistance.
    b. Hazard Mitigation Assistance.
    c. Individual and Business Assistance.
    d. Community Assistance.

26. What federal agency makes a recommendation to the President when a federal declaration or disaster is being considered? _____.

    a. The National Security Commission.
    b. The National Weather Service.
    c. The Federal Emergency Management Agency.
    d. The Department of Agriculture.

27. The Governor must request Federal disaster assistance before a Presidential disaster declaration can be granted.

   a. True
   b. False

28. Household items, furnishings, and appliances damaged in a disaster might be replaced or repaired through the _____.

   a. Individuals and Households Program.
   b. Legal services program.
   c. Social Security fund.
   d. Public assistance fund.

29. Certain privately owned facilities, such as airports and hospitals, might receive federal assistance for repairs if there were a federal declaration of disaster.

   a. True
   b. False

30. The organization that provides flood insurance to qualifying communities who choose to participate is the _____.

   a. National Guard.
   b. Department of Labor.
   c. National Flood Insurance Program.
   d. Flood Safety Agency.

31. In a Presidentially declared disaster, disaster victims can apply for assistance at _____.

   a. The Federal Disaster Office.
   b. The Disaster Recovery Center or National Processing Service Center.
   c. The Emergency Management Office.
   d. Any Red Cross-operated shelter.

32. In a Presidentially declared disaster, the person responsible for coordinating the overall disaster recovery effort at the federal level is the

   _____.

   a. Mayor of the affected community.
   b. Federal Coordinating Officer.
   c. State Emergency Manager.
   d. Governor.

33. The Stafford Act does not explicitly prohibit receiving more than one payment for the same loss in a Presidentially declared disaster.

   a. True
   b. False

34. An applicant for individual assistance is notified of FEMA's decision concerning that assistance by _____.

   a. Phone
   b. Mail
   c. E-mail
   d. A FEMA representative

35. Applicants for disaster relief may be requested to supply proof of residence and information on insurance coverage.

   a. True
   b. False

36. Applicants for federal disaster relief in a Presidentially declared disaster can expect to receive an assistance check at the time of application.

   a. True
   b. False

37. A National Flood Insurance policyholder does not have to wait for a Presidential disaster declaration before filing an insurance claim for flood damage.

   a. True
   b. False

38. In a Presidentially declared disaster, federal disaster relief and recovery efforts are coordinated at what site? _____.

    a. Disaster Recovery Center.
    b. Federal Emergency Management Office.
    c. Disaster Field Office.
    d. Stafford Office.

39. Government officials at all levels generally use public information officers (PIOs) to keep the public apprised of developments in the recovery effort.

    a. True
    b. False

40. The best source of information on a particular community's disaster history is _____.

    a. The community's local emergency management agency.
    b. The National Guard.
    c. Local elected officials.
    d. The local police department.

41. Insurance claims may be expedited if an inventory of possessions, supported by photographs and/or video, has been prepared in advance.

    a. True
    b. False

42. In the event of a disaster affecting the home, such as fire, family members should plan to meet _____.

    a. In the basement.
    b. Within a few feet of the primary entrance.
    c. At a relative's home in another town.
    d. At a pre-designated location at a safe distance from the home.

43. Household members should have enough supplies to take care of themselves for at least _____ if a disaster occurs.

    a. 12 hours.
    b. 24 hours.
    c. 48 hours.
    d. 72 hours.

44. If an evacuation were ordered because of a disaster, notice would always be given at least one day in advance.

    a. True
    b. False

45. Once the utilities have been turned off, only a service representative or other knowledgeable person should turn them back on.

    a. True
    b. False

46. Smoke detector batteries should be changed how often? _____.

    a. Every 18 months.
    b. At least once a year.
    c. Every other year.
    d. Whenever you think of it.

47. Every member of a household—even children—should know how to summon help in an emergency.

    a. True
    b. False

48. Disaster victims can expedite the process of applying for assistance by being prepared to provide _____.

    a. The address and telephone number where they can be reached.
    b. An inventory of damage.
    c. Insurance information.
    d. All of the above.

49. Evacuations are extremely rare in the United States—occurring less than five times annually.

    a. True
    b. False

50. It is important to plan alternate escape routes that could be used in the event of fire.

    a. True
    b. False